Manufacturing and Novel Applications of Multilayer Polymer Films

PLASTICS DESIGN LIBRARY (PDL)
PDL HANDBOOK SERIES
Series Editor: Sina Ebnesajjad, PhD (sina@FluoroConsultants.com)
President, FluoroConsultants Group, LLC
Chadds Ford, PA, USA
www.FluoroConsultants.com

The **PDL Handbook Series** is aimed at a wide range of engineers and other professionals working in the plastics industry, and related sectors using plastics and adhesives.

PDL is a series of data books, reference works and practical guides covering plastics engineering, applications, processing, and manufacturing, and applied aspects of polymer science, elastomers and adhesives.

Recent titles in the series

Biopolymers: Processing and Products, Michael Niaounakis (ISBN: 9780323266987)
Biopolymers: Reuse, Recycling, and Disposal, Michael Niaounakis (ISBN: 9781455731459)
Carbon Nanotube Reinforced Composites, Marcio Loos (ISBN: 9781455731954)
Extrusion, 2e, John Wagner & Eldridge Mount (ISBN: 9781437734812)
Fluoroplastics, Volume 1, 2e, Sina Ebnesajjad (ISBN: 9781455731992)
Handbook of Biopolymers and Biodegradable Plastics, Sina Ebnesajjad (ISBN: 9781455728343)
Handbook of Molded Part Shrinkage and Warpage, Jerry Fischer (ISBN: 9781455725977)
Handbook of Polymer Applications in Medicine and Medical Devices, Kayvon Modjarrad & Sina Ebnesajjad (ISBN: 9780323228053)
Handbook of Thermoplastic Elastomers, Jiri G Drobny (ISBN: 9780323221368)
Handbook of Thermoset Plastics, 2e, Hanna Dodiuk & Sidney Goodman (ISBN: 9781455731077)
High Performance Polymers, 2e, Johannes Karl Fink (ISBN: 9780323312226)
Introduction to Fluoropolymers, Sina Ebnesajjad (ISBN: 9781455774425)
Ionizing Radiation and Polymers, Jiri G Drobny (ISBN: 9781455778812)
Manufacturing Flexible Packaging, Thomas Dunn (ISBN: 9780323264365)
Plastic Films in Food Packaging, Sina Ebnesajjad (ISBN: 9781455731121)
Plastics in Medical Devices, 2e, Vinny Sastri (ISBN: 9781455732012)
Polylactic Acid, Rahmat et. al. (ISBN: 9781437744590)
Polyvinyl Fluoride, Sina Ebnesajjad (ISBN: 9781455778850)
Reactive Polymers, 2e, Johannes Karl Fink (ISBN: 9781455731497)
The Effect of Creep and Other Time Related Factors on Plastics and Elastomers, 3e, Laurence McKeen (ISBN: 9780323353137)
The Effect of Long Term Thermal Exposure on Plastics and Elastomers, Laurence McKeen (ISBN: 9780323221085)
The Effect of Sterilization on Plastics and Elastomers, 3e, Laurence McKeen (ISBN: 9781455725984)
The Effect of Temperature and Other Factors on Plastics and Elastomers, 3e, Laurence McKeen (ISBN: 9780323310161)
The Effect of UV Light and Weather on Plastics and Elastomers, 3e, Laurence McKeen (ISBN: 9781455728510)
Thermoforming of Single and Multilayer Laminates, Ali Ashter (ISBN: 9781455731725)
Thermoplastics and Thermoplastic Composites, 2e, Michel Biron (ISBN: 9781455778980)
Thermosets and Composites, 2e, Michel Biron (ISBN: 9781455731244)

To submit a new book proposal for the series, or place an order, please contact David Jackson, Acquisitions Editor
david.jackson@elsevier.com

Manufacturing and Novel Applications of Multilayer Polymer Films

Deepak Langhe
Michael Ponting

PolymerPlus LLC, Valley View, Ohio, USA

ELSEVIER

Amsterdam • Boston • Heidelberg • London
New York • Oxford • Paris • San Diego
San Francisco • Singapore • Sydney • Tokyo
William Andrew is an Imprint of Elsevier

William Andrew is an imprint of Elsevier
The Boulevard, Langford Lane, Kidlington, Oxford, OX5 1GB, UK
50 Hampshire Street, 5th Floor, Cambridge, MA 02139, USA

British Library Cataloguing-in-Publication Data
A catalogue record for this book is available from the British Library

Library of Congress Cataloging-in-Publication Data
A catalog record for this book is available from the Library of Congress

ISBN: 978-0-323-37125-4

Publisher: Matthew Deans
Acquisition Editor: David Jackson
Editorial Project Managers: Peter Gane and Zara Preston
Production Project Manager: Julie-Ann Stansfield
Designer: Mark Rogers

Typeset by Thomson Digital

Contents

Preface

Modern day technology automation and advanced processing equipment have enabled a more rapid product development culture in industrial as well as academic research and development arenas. Focused efforts to manufacture products that are optimized as cheaper, greener, and higher performing, have pushed materials and technology development to interdisciplinary science and engineering solutions. The highly creative nature of these products requires processing or production tools capable of the same. Originally created as a specialty advanced polymer processing technique for novelty consumer packaging, micro- and nanolayer coextrusion technology has become an increasingly important solution for product manufacturing in the polymer and composite film and sheet fields.

Layered polymeric and composite products have been utilized as a developmental tool since the paper mills of the late 1800s. Today, layered products are heavily integrated into all aspects of everyday life; from cheese wrappers to tennis shoes, and shatter proof car windshields to high performance optical camera lenses. In support of the desire to reduce material consumption, cost, and complexity in products, additional functionality, and compatibility with next generation materials is required. Over the last twenty years, the trend in polymeric film and sheet development has taken place to increase the number of materials. Today products can contain up to 11 unique materials and layers. As the polymer industry continues to accept products comprised of 10, 20, or even 30 layers, the state-of-the-art products are again approaching the micro- and nanolayer thicknesses originally invented and researched back in the 1960s.

It is said that those who do not know history are doomed to repeat it. One mission of this text was to catalogue the 50+ years of micro- and nanolayered polymeric research and development, while inspiring continued product evolution and creation of next generation applications, and innovations. Many emerging applications and recent advances, along with a wide spectrum of properties, have been discussed in this book. Arbitrary control of molecules on the nanometer-sized scale via a continuous, scalable, and industrially compatible process, is a very rare tool that the authors hope will continue to be exploited in the near future of polymer and material scientists and engineers.

Deepak Langhe
Michael Ponting

1 Introduction to Multilayered Films

1.1 Introduction

Development of polymer blends and composites rely on the synergistic combination of two or more materials to achieve enhanced properties. Frequently, it is necessary to use blend or composite structures because single polymer components may not meet the end use requirements. The final product properties are strongly dependent on several blend and material parameters including formulation, morphology, processing history, interactions between the components, and other intrinsic properties of the individual components. Structure–property relationships for polymer blends and composites have been reviewed and discussed previously [1]. Other approaches such as copolymerization, formulations with processing aids, compatibilizers, impact modifiers, and fillers have been employed to combine the polymers more effectively. One approach to combine the two or more polymers is coextrusion process that combines two or more polymers as a layered structure to achieve wide range of property improvements. Similar to blending, the coextrusion process also relies on achieving synergistic effects by combining polymers at different scales, however, adds an additional structural design freedom that offers control over the size and location of the constituent material domains.

For more than six decades, layered composites have demonstrated their importance in the commercial marketplace. Unlike many earlier technologies, which required multiple manufacturing steps to create multicomponent polymer-layered composites, the coextrusion process produces multilayer films using a single continuous processing step. This advantage reduces manufacturing complexities associated with other processing techniques such as multistep lamination and coating, which require separate handling of individual layers or plies. In contrast, the coextrusion processing allows simultaneous combinations of two or more polymer materials inside a die or feedblock to produce layered structures, thus eliminating the need for any postprocessing or repeated fabrication steps. In 1950s, many thermoplastic film applications were developed with only a few layers. The emergence of an advanced coextrusion processing technique enabling the fabrication of hundreds of thousands of multilayers in coextruded articles has recently led to capabilities and new breakthroughs in material properties and development. The ability to order multiple polymer materials on the micro- or nanometer size scale have led to improvements in many properties and

Manufacturing and Novel Applications of Multilayer Polymer Films
http://dx.doi.org/10.1016/B978-0-323-37125-4.00001-0

the discovery of novel material properties that comprise the discussed technology and case studies of this text.

Many conventional packaging applications may contain only a few polymer layers (typically between two and seventeen) with diverse properties. Typical layered structure can contain as many as five to seven different polymers with varied material properties such as barrier improvement, chemical resistance, and sealability properties. In these applications, the individual layer thicknesses ranged from the millimeter size scale down to few hundreds of micrometers. In 1960s, the Dow Chemical Company developed the technology to combine two polymers as hundreds of layers by coextrusion process and layer-multiplying die elements [2]. This technology allowed production of films composed of hundreds or thousands of layers with the individual layer thicknesses down to few tens of nanometers.

The objective of this book is to discuss the properties and applications of multilayered films especially related to nanolayered films. In general, many novel applications and recent advancements of coextruded structures are discussed.

1.2 Coextrusion Processing Techniques

Coextruded structures are produced by various techniques, mainly blown film and cast film processes. Blown film process is one of the most commonly used manufacturing technologies in plastic film production for consumer food and medical packaging. In this process, a tubular coextrusion process uses formation of concentric layers in the annular die. As the melt extrudate exits the die, it is expanded with air pressure and subsequently cooled to form the final film. Tubular coextrusion process uses several dies with different designs including single manifold, multimanifolds, or toroidal distribution manifolds. Typical design requirements are formation of concentric layers in annular die with mandrel and outside die ring. In the single manifold design, two polymer melts are arranged as concentric melt streams to produce two-layered annular extrudate [3]. In more commonly used multimanifold design, the individual manifolds with different polymer materials are distributed concentrically, which are joined prior to melt stream exit. Several concentric manifolds can be present around the mandrel. Although earlier designs produced two- or three-layered films, recent advancements produce dies with seven or more layer [4]. Toroidal manifolds cause sequential addition of polymer materials to produce desired number of layers. The manifolds are typically designed polymer-specific to allow uniform circumferential distribution of the polymer melt. As the number of layers increases, the manifold designs can become very complex. To create hundreds of layers, an annular die with rotating elements was designed [5]. The extruded polymers arrange into alternating, radially extending layers in the die and deformed into thin spirals by rotating elements. However, this method does not necessarily force polymers into layered structures like other techniques. Another tubular blown film technique uses stackable plate die. In this technique, each polymer layer is formed into a tube in a single plate. Multiple such plates can be arranged sequentially to achieve desired number of layers.

In the cast film coextrusion process, a flat die geometry allows for polymer material extrusion through a wide but narrow slit. Two types of die geometries used in this process are, multimanifold dies and the single manifold die with a feedblock. Multimanifold dies with individual manifolds for each separate polymer material layer extend across the width of the die to combine the melt streams to produce a multilayer sheet. In externally combining dies, the polymer layers are combined at the roll nip and typically limited to two-layer coextrusion in a lamination process. Another layered materials processing technique, a layering feedblock technique, combines two or more polymers melt stream ahead of the single-manifold die inlet. The feedblock design typically consists of a modular feedblock with ability to combine three polymers to produce a five-layer structure. The polymer melt stream with layers can be arranged into different shapes (round, square, rectangular) prior to die inlet [6]. Polymer melt from each extruder can be subdivided into many layers as desired. As the polymer melt spreads through the die, reduction in layer thickness is achieved. Earlier innovations produced five- or six-layered films and sheets in several layer arrangements. The feedblock can be connected with layer-multiplying elements or the interfacial surface generating devices to create hundreds of layers, which is also the basis of technologies discussed in this book. The flexibility of the feedblock method has made it one of the most commonly used coextrusion processing technique. Alternative coextrusion feedblock technologies can employ movable vane partitions to process polymers with large viscosity mismatches. Another alternative processing technique utilizes a combination of feedblocks and multimanifold dies to produce film systems with more than three layers comprised of five to nine unique polymer materials. The flexibility of the feedblock technique has made it one of the most commonly used coextrusion process technology [4].

1.3 Toward Hundreds of Layers

Most of the techniques summarized previously are suitable for production of multilayered films with only a few layers, which still offer many improved properties in packaging and industrial applications. However, advancements in the development of multilayered films with hundreds of layers have produced materials with many unique properties. One of the approaches to produce multilayered films with hundreds of layers was developed by Schrenk [7]. A feedblock design included combining polymer melts from two polymers from the opposite side of the feedblock, subdivided into substreams and then interdigitated to produce multilayered structures. In a recent invention, Dooley et al. developed multilayer film structures with annular profile having large number of layers and a reduced number of manifolds using a modified crosshead-style blown film die [8]. In this method a multilayer flow stream with four or more layers was fed into a distribution manifold to form an annular flow stream and then split into two flow streams moving in opposite directions and overlap to form annular profile.

In a more commonly used approach, interfacial surface generators, also referred to as layer-multiplying dies, are utilized to increase the number of film layers by sequentially splitting a two or three-layered polymer melt streams in half, followed by a stacking-and-spreading step to double the number of layers [9,10]. As this process "forces" polymers to flow into the desired layered geometry, it is also known as a "forced assembly" coextrusion process. The interfacial surface generator, as shown in Figure 1.1, can split the polymer melt and stack again to increase the number of layers. For example, a two-layer starting structure would produce 2^{n+1} layers using "n" number of layer multiplier dies placed in a series. This assembly will produce film with alternating layers of two polymers with a repeating

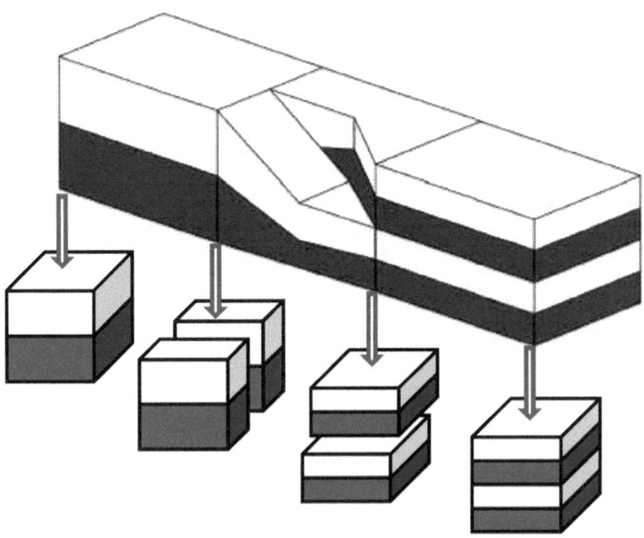

Figure 1.1 Layer multiplication starting with two layers of equal thickness; and the layer pattern produced in different sections of an interfacial surface generator.

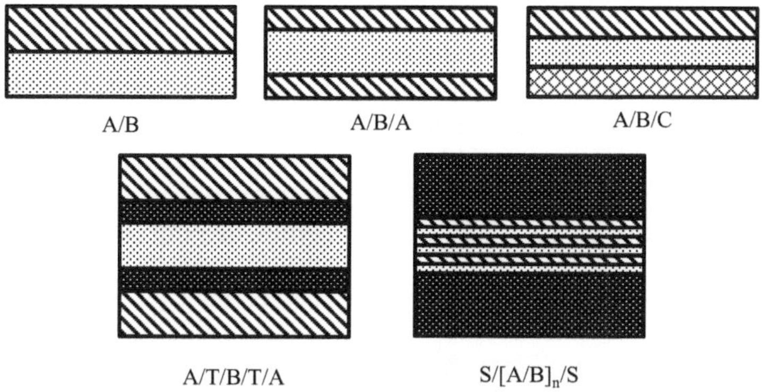

Figure 1.2 Different types of layer-structure morphologies produced using coextrusion process. A, B, and C represent three different types of polymers. Top row represents the repeating units that can be produced using layer multiplication process. It is also possible to extend this approach beyond three-layer repeating units to five or more. Bottom row images represent addition of tie layer (T) and skin layer (S) materials.

layer sequence of (AB), where A and B represent two different polymers. Nanolayered structures with more than 8000 layers have been produced in a laboratory set-up [11]. The number of layers, the melt feed ratio of the polymers, and the final film thickness can be independently controlled to achieve cost effective, custom product formulations with layer thicknesses in micro- or nanoscale. The process is also very versatile as it allows addition of a third polymer (polymer C), followed by layer-multiplication process to create layered structure with ABC repeating unit. Most often, this third component is added as a tie-layer polymer between polymers A and B to improve their compatibility or adhesion. This type of layer configuration produces a repeating unit of (ATBTA). Another process modification allows adding skin layers after the layer multiplication dies prior to the cast film die. Examples of some commonly produced layered structures are shown schematically in Figure 1.2. The layer structure integrity and thickness uniformity is dependent on the viscosity ratio of the components as discussed in Chapter 2. Advanced work in numerical simulations for velocity counters and flow patterns have been performed and summarized by Dooley [4]. In another approach, Cloeren developed feedblocks to split the melt flow from two polymers into different streams to create layered structures, which are melt laminated together within the feedblock to produce multilayered composites with a few hundred layers using a conventional cast film die [12].

Most of the advanced multilayer film properties and potential applications discussed in this book are based on layer multiplication coextrusion processing technique and can be extended to products made using feed-block technology. The fundamental understanding of the structure–property relationships on micro- and nanolayered structures is important in all processing techniques. The multilayer coextrusion processing and the factors affecting coextruded structures processing and uniformity are discussed in Chapter 2. In this book, the coextrusion process refers to multilayered structures with number of layers varying from tens to thousands.

1.4 Multilayer Film Applications

Many unique properties of multilayered films are reported in the literature. For example, iridescent multilayered films and articles were made by combining polymers with different refractive indices. This phenomenon was further explored to develop high reflective multilayered films for infrared, visible, and ultraviolet light reflection applications [13]. Other applications as optical films include brightness enhancing optical films or reflective polarizers [14] and films for window glass for improvement toughness, blocking UV light and automotive glazing. For example, 3M's Ultra Series of multilayered films with up to 42 layers is used for its excellent impact properties. In other example, polypropylene/polystyrene multilayered films showed improved water barrier and elongation properties in crumpled samples as compared to polystyrene films [13,15,16]. Other examples of multilayered films with improved mechanical properties include multilayered sheets of polycarbonate (PC) and styrene-acrylonitrile (SAN), which showed drawing of SAN layers in ductile manner with suppressed crazing [17]. Increased number of layers showed many beneficial effects such as improved brittle to ductile transition and impact properties at lesser PC content in the multilayered composite. A 10-fold increase in the mechanical toughness and impact strength was observed as the layer thickness decreased from tens of micrometers to microscale. Recent innovations are also discussed concerning the effect of submicrometer scale layer thicknesses on deformation of polymer nanolayers [18]. The key improvements in barrier and mechanical properties allowed use of multilayered films in applications such as flexible barrier materials or bladders for shoes [19], autoclavable barrier films for medical applications [20], and barrier packaging material for food containers [21]. As discussed in these applications and examples later, coextrusion process has been used to process wide range of thermoplastic polymers to create low cost as well as high value added products.

1.5 Multilayer Film Properties

During the process of micro- or nanolayer coextrusion, the intrafilm layer thicknesses can be reduced to the size scale of individual polymer molecule dimensions resulting in polymer interfacial regions and properties become significantly large contributors to the overall macrofilm properties. Many properties such as crystallization, adhesion, interfacial interactions, and interdiffusion are affected as the interface to volume ratio increased. The layer multiplication coextrusion process allows manipulating the relative composition of the polymers as well as final film thickness to produce multilayered films with micro- and nanoscale, which are used to investigate the complex behavior of multilayered structures. In recent years, with advanced material characterization techniques, fundamental understanding of the structure–property relationships in the layered films has shown dramatic effect of layer thickness on various properties. Many such properties of the multilayered films are summarized in subsequent sections and discussed in detail in subsequent chapters of the book.

1.5.1 Gas Barrier Properties

Multilayered films with micro- and nanoscale confinement demonstrated dramatic changes in the crystallization behavior and impacted the gas transport properties [22]. The confined crystallization of polymers led to formation of oriented lamellae in nanolayered films, which showed two to three orders of improvement in the gas barrier properties. The oriented lamellae increased the tortuosity to gas diffusion pathway with increasing the lamellar orientation thereby improving the barrier performance. As the layer thicknesses approached few nanometers, lamellar morphologies in polyethylene oxide (PEO) and polycaprolactone (PCL) confined layers resembled "single crystal" structures. This phenomena of confined crystallization was also demonstrated in many other polymers such as syndiotactic polypropylene (sPP), polyvinylidene fluoride (PVDF), high-density polyethylene (HDPE) and poly(4-methylpentene-1) (P4MP1) in layered films and under appropriate thermal treatment showed changes in the gas and water barrier properties. Furthermore, the "confining" polymers also played an important role in controlling the crystal orientation of the "confined" polymer. Although multilayered films have demonstrated improved gas barrier properties previously, to achieve a two to three orders of magnitude improvement, the layer thicknesses between few tens and a few hundred nanometers were required. Therefore, such barrier improvement is not possible under conventional coextruded films with only a few layers. Layer multiplication coextrusion processing and feedblock technology allowed fabrication of layers with hundreds of layers and film thicknesses in the range of few micrometers as necessary for many packaging applications. Furthermore, the multilayered structures were also incorporated in biaxially oriented PP and polyethylene terephthalate (PET) as skin layers allowed biaxial stretching of multilayered films to produce optically clear and mechanically durable films. The effects of confined nanolayers and structure–property relationships of the layered films are discussed in Chapter 3.

1.5.2 Adhesion Properties

As multilayer coextrusion process creates large number of interfaces, the higher ratio of interfaces to volume allowed probing adhesion properties of polymers. Interfacial adhesion is critical in defining the final mechanical properties of the composites. For example, in PC/SAN composites, the relative thicknesses of PC and SAN polymers changed the layer delamination mechanism [23]. The multilayer coextrusion process flexibility allowed fabrication of multilayered films with a tie layer polymer, which act as a glue layer or a compatibilizer. For example, PP/HDPE multilayered films coextruded with olefinic block copolymers (OBCs) and ethylene–octene copolymers (EOs) improved the adhesion between PP and HDPE [24]. This approach was also used to estimate the delamination toughness and effect of chain architecture on adhesion properties. Multilayer films offer an opportunity to investigate compatibilizers and effectively use them in multilayers and blends.

1.5.3 Mechanical Properties

Many multilayered composites showed enhanced mechanical properties with increasing number of layers, even when the relative composition of the polymers was kept constant. For example, in PC/SAN and PC/PMMA layered composites, the deformation mechanism of the SAN and PMMA layers changed from cavitation and craze opening to shear yielding as the layer thickness decreased. PC/SAN composites also showed a 10-fold enhancement in the impact strength as the layer thickness changed from macroscale to microscale [25]. Improved impact properties and fatigue resistance were also observed in these multilayered composites. The fatigue resistance properties can be useful in high-pressure pipe applications. Further investigation of deformation mechanism in nanoconfined PEO lamellae offered insights into deformation of orientated lamellae [26]. Wide range of multilayer film systems with filled polymers, gradient layer thicknesses, and block copolymers have also been investigated. Investigations related to adhesion and mechanical properties can effectively be used to design films for packaging applications with film-down gauging possibilities.

1.5.4 Interphase Materials

Properties of the multilayered composites are strongly dependent on the interphase formation and interdiffusion of polymers during coextrusion processing. To explore the interphase properties, multilayered films with two completely immiscible polymer pairs were extruded with variable number of layers [27]. For example, multilayered films of PC and polymethyl methacrylate (PMMA) extruded with increasing number of layers (or decreasing layer thickness) exhibited a merging of glass transition temperature as layer thicknesses decreased below 100 nm. At layer thicknesses of 10 nm or less, interphase material indicated by convergence of the glass transition temperatures were reported. In fact, the layer thicknesses were close to the estimated interphase thickness of 9 nm for PC/PMMA system calculated using Helfand and coworker's theoretical prediction [28]. Multilayer coextrusion process allowed probing of interphase characteristics, which are otherwise not possible to access using conventional techniques. The fundamental understanding of the interphase materials has wide implications in designing multilayered films as well as developing blends, and composite materials.

1.5.5 Interdiffusion in Polymers

Movement of the polymer chains across the interface, measured as the mutual diffusion coefficient, defines the interdiffusion properties of polymers. Multilayer coextrusion provided an opportunity to observe interdiffusion in the multilayered structures by coextruding two miscible polymers [29]. Particularly, the laminar flow conditions in the coextrusion processing combined polymers in the layer multipliers by producing large interfacial area without complete mixing. The symmetrical structures and uniform nature of the multilayered structures allowed investigation of interdiffusion in many polymer systems such as PC/copolyesters, Nylon/EVOH and different molecular weight polyethylenes. Different parameters such as number of layers, temperature and layer thicknesses were investigated to model the composition profiles in the layer composites.

1.5.6 Optical Properties

Some of the early innovations and many commercial applications of multilayered films are optical, iridescent, and reflective films. Multilayer films are used in wide range of applications ranging from decorative films to enhanced optical displays. Combination of the polymers with different refractive indices allowed creation of optically reflective films for infrared, visible, and ultraviolet light reflection capabilities. Fabrication of narrowband one-dimensional photonic crystals was achieved in multilayered films due to periodicity in the refractive index in the multilayered films [30]. The flexibility of the coextrusion process allowed fabrication of elastomeric photonic crystals with potential applications as strain sensors, tunable optical filters, and mirrors [31]. Other applications such as brightness enhancing optical films have also been demonstrated using multilayered films. Multilayered polymer films with a designed photonic band gap in combination with fluorescent dyes were used to fabricate distributed Bragg reflector and distributed feedback lasers. Optical films with dopant molecules demonstrated application in optical data storage.

1.5.7 Dielectric Properties

Multilayered films of PVDF homopolymer or copolymers and PC or PET polymers demonstrated increased energy density, breakdown strength, and low losses [32]. Coextrusion processing of high breakdown strength and high dielectric constant polymers showed synergistic effect in the layered films. The dielectric breakdown mechanism revealed importance of interfaces in improving the breakdown properties. It is interesting to note that the polymer capacitor film technology has not changed significantly in last two decades. Multilayered dielectric films with increased energy density properties can be used to produce compact, high energy density capacitors.

1.6 Novel Applications

Many property improvements suggest a broad application space for multilayered structures beyond conventional packaging and optical film applications. In addition to the properties and applications discussed earlier, many other novel uses of multilayered films and composites were investigated.

Advanced blends and composites: Multilayered films have been used as blends at appropriate processing temperatures. Coextrusion process was used as a tool to create high aspect ratio microplatelets using multilayered films [33]. For example, using PP/PA66 multilayered system, microplatelets of PA66 in PP matrix were achieved by selectively melting PP layers to form the matrix, while PA66 layers remained solid as dispersed layers.

The difference in the melting temperatures of PP (T_m = 163°C) and PA66 (T_m = 262°C) allowed processing of the pelletized multilayer structures at temperatures above PP melting temperature and below PA66 melting temperatures. The morphology of molded plaques confirmed PA66 microplatelets dispersion in PP matrix. The resulting material showed improvement in the gas barrier properties as compared to the conventional melt blends. Structural models revealed that the tortuosity of the gas diffusion pathway increased significantly due to microplatelets.

Foam/film structures: Different types of multilayer foam/film structures including PP, PE, ethylene–octene (EO) copolymer, cross-linked polyolefin elastomers, and polyolefin blends were developed using coextrusion process [34]. Although, it is inherently difficult to control the cell size in polymers such as PP and ethylene–octane copolymer, improved foaming efficiency and controlled cell size were achieved in multilayered composites. In the layered foam/film structures, one of the polymers contained a chemical blowing agent and a nucleating agent, which decomposed during processing to create foam layer. The second polymer remained intact as film layers separating the foam layers. Examples of PP-based multilayer foam/film structures replicated the microcellular structure of the natural cork. The foam/film composites can find uses in many applications including packaging, automotive, construction, sports, etc.

Porous materials: Multilayered films have also been demonstrated as active gas separation membranes with selective polymer and porous support layers. For example, multilayered films of poly(ether block amide) (PEBA) elastomers with filled PP layers were oriented to create porous PP layers, while PEBA layers remained intact [35]. The coextruded structures demonstrated high CO_2/O_2 selectivity and moderate permeability. The approach demonstrated possibility of creating polymer separation membranes with porous supports as membranes or modified atmosphere-packaging materials. In another approach, stretched multilayered films of plastic/elastic polymers created corrugated structures for applications as breathable films [36].

Gradient structures: Coextrusion processing also explored the production of multilayered films with gradient layer thickness distribution for wide range of optical applications [37]. Using custom-designed layer multiplication dies to create gradient structures or custom layer thickness configurations, it is possible to design optical filters. The flexibility of this process was also demonstrated in creating gradient film/foam composites.

Shape memory polymers: Combining two or more polymers with different thermal and mechanical properties, the fabrication of shape memory polymers was demonstrated in polyurethane (PU) and polycaprolactone (PCL) multilayered films [38]. Ability of multilayered films to change shape under external temperature stimulus was demonstrated in PU/PCL films.

Multilayer micro- and nanofibers: Many different approaches have been investigated to create fibrillar morphology using multilayered films. Using side by side orifices in a cast film die, bicomponent fibers containing alternating layers of two polymers were demonstrated [39]. Alternatively, the coextrusion process was modified as "two-dimensional" polymer melt multiplication technique to create films with vertical layer stacks separated by horizontal layers [40]. The structure resembled hundreds of fibers in a polymer film, which were subsequently separated by hydro-entanglement technique.

Filled composites: Different examples of filled microlayer composites have created brick-wall type microstructures in the multilayer composites, which showed improved barrier and mechanical properties [41]. Interestingly, some natural systems like nacre consisted of layered structure with aragonite inorganic bricks in an organic matrix, which exhibited excellent fracture toughness as compared to monolithic aragonites [42]. Other examples of filled multilayered composites with oriented and aligned filler particles were also discussed [43].

1.7 Summary

From the wide range of reported layered structure property improvements and applications, the coextrusion process is a unique and novel processing approach of great commercial significance. The coextrusion process has evolved from its early days of two- or three-layered systems for packaging applications to films and products comprised of hundreds of layers, which demonstrate novel optical, mechanical, or transport applications. Better understanding of structure and properties of the layered polymers has expanded the applications space for multilayered composites beyond packaging to include energy storage, optical devices, and sensors. Using layer multiplication process and feedblock technology, producing hundreds to thousands of layers is accessible in commercial production facilities. Furthermore, this approach offers unique opportunities to probe nanoscale material interactions via easily handled macrofilms, hundreds of micrometers thick, can be analyzed utilizing conventional polymer and material science analytical equipment and techniques. Nanoscale phenomena studies including material interphase composition, interdiffusion, confined crystallization, optical behavior, orientation, homogeneous and heterogeneous nucleation have been successfully probed via the micro- and nanolayer coextrusion method. In this book, we highlight the current capabilities of microlayer coextrusion processing, state-of-the-art research into nanolayered polymer systems and composites and report on the novel applications and devices currently enabled. Finally, a look toward the future of nanolayered coextrusion and its potential development areas is discussed.

Note: "Polymer A/Polymer B" nomenclature represents a multilayered film with alternating layers of Polymer A and Polymer B. For example, PC/PMMA film denotes a multilayered film with alternating layers of PC and PMMA polymers. All the compositions of the film component are volume percentages, unless mentioned otherwise.

References

[1] (a) D.R. Paul, C.B. Bucknall, Polymer Blends Volume 1: Formulation, A Wiley-Interscience Publication, NY, 2000. (b) D.R. Paul, C.B. Bucknall, Polymer Blends Volume 1: Performance, A Wiley-Interscience Publication, NY, 2000.
[2] R.E. Harder, US Patent No. 3,195,865, 1965.
[3] (a) US Patent No. 3,223,761, 1965. (b) J.E. Johnson, Plast. Technol. 22 (1976) 45–48.
[4] J. Dooley, Viscoelastic flow effects in multilayer polymer coextrusion. Ph.D. Thesis, 2002.
[5] (a) W.J. Schrenk, T. Alfrey Jr., SPE J. 29 (1973) 38. (b) W.J. Schrenk, T. Alfrey Jr., SPE J. 29 (1973) 43.
[6] L.M. Thomka, W.J. Schrenk, Mod. Plast. 49 (4) (1972) 62–64.
[7] W.J. Schrenk, US Patent 3,884,606, 1975.
[8] J. Dooley, J.M. Robacki, M.A. Barger, R.E. Wrisley, S.L. Crabtree, C.L. Pavlicek, US Patent No. 2010/0,215,879, 2010.
[9] W.J. Schrenk, R.K. Shastri, R.E. Ayres, US Patent No. 5,094,793, 1992.
[10] W.J. Schrenk, R.K. Shastri, R.E. Ayres, US Patent No. 5,094,788, 1992.
[11] T.E. Bernal, A. Ranade, A. Hiltner, E. Baer, in: G.H. Michler, F. Balta-Calleja (Eds.), Mechanical Properties of Polymers Based on Nanostructure and Morphology, Taylor & Francis, Florida, 2005, pp. 629–682 (Chapter 15).
[12] (a) P.F. Cloren, US Patent Application 2005/0,029,691 A1, 2005. (b) P. Cloeren, US Patent No. 2003/0,201,565 A1, 2003.
[13] T. Alfrey, Jr., W.J. Schrenk, US Patent No. 3,711,176, 1973.
[14] R.C. Allen, L.W. Carlson, A.J. Ouderkirk, A.L. Kotz, T.J. Nevitt, C.A. Stover, B. Manumdar, US Patent No. 6,111,696, 1996.
[15] W.J. Schrenk, D.S. Chisholm, K.J. Cleereman, T. Alfrey, Jr., US Patent No. 3,576,707, 1969.
[16] W.J. Schrenk, T. Alfrey Jr., Polym. Eng. Sci. 9 (6) (1969) 393–399.
[17] (a) M. Ma, K. Vijayan, A. Hiltner, E. Baer, J. Mat. Sci. 25 (1990) 2039–2046. (b) D. Haderski, K. Sung, J. Im, A. Hiltner, E. Baer, J. Appl. Polym. Sci. 52 (1994) 121–133.

[18] R. Adhikari, V. Seydewitz, K. Loschner, G.H. Michler, A. Hiltner, E. Baer, Macromol. Symp. 290 (2010) 156–165.
[19] P.H. Mitchell, J.C. Sell, Jr., H.W. Bonk, US Patent No. 5,713,141, 1998.
[20] M.O. Chang, M.T. Ling, Y.S. Ding, US Patent No. 8,097,346 B2, 2012.
[21] A.M. Chuprevich, M. Bentmar, US Patent Application No. US 2007/0,269,622 A1, 2006.
[22] (a) H.P. Wang, J.K. Keum, A. Hiltner, E. Baer, B. Freeman, A. Rozanski, A. Galeski, Science 323 (2009) 757–761. (b) M. Ponting, Y. Lin, J.K. Keum, A. Hiltner, E. Baer, Macromolecules 43 (2010) 8619–8627. (c) D.S. Langhe, A. Hiltner, E. Baer, Polymer 52 (25) (2011) 5879–5889. (d) M. Mackey, L. Flandin, A. Hiltner, E. Baer, J. Polym. Sci. Part B Polym. Phys. 49 (2011) 1750–1761. (e) G. Zhang, P.C. Lee, S. Jenkins, J. Dooley, E. Baer, Polymer 55 (2014) 663–672.
[23] T. Ebeling, A. Hiltner, E. Baer, J. Appl. Polym. Sci. 68 (1998) 793–805.
[24] (a) A.R. Kamdar, R.K. Ayyar, B.C. Poon, G.R. Merchand, A. Hiltner, E. Baer, Polymer 50 (2009) 3319–3328. (b) P. Dias, Y.J. Lin, B. Poon, H.Y. Chen, A. Hiltner, E. Baer, Polymer 49 (2008) 2937–2946.
[25] J. Im, A. Hiltner, E. Baer, High performance polymers, in: E. Baer, A. Moet (Eds.), Microlayer Composites, Hanser, New York, 1991, pp. 175–198.
[26] C. Lai, R. Ayyar, A. Hiltner, E. Baer, Polymer 51 (2010) 1820–1829.
[27] (a) R.Y.F. Liu, Y. Jin, A. Hiltner, E. Baer, Macromol. Rapid Commun. 24 (16) (2003) 943–948. (b) R.Y.F. Liu, A.P. Ranade, H.P. Wang, T.E. Bernal-Lara, A. Hiltner, E. Baer, Macromolecules 38 (2005) 10721–10727.
[28] (a) E. Helfand, A.M. Sapse, J. Chem. Phys. 62 (1975) 1327–1331. (b) E. Helfand, Y. Tagami, J. Polym. Sci. Polym. Lett. 9 (1971) 741–746. (c) E. Helfand, Y. Tagami, J. Chem. Phys. 56 (1972) 3592–3601.
[29] (a) G. Pollock, S. Nazarenko, A. Hiltner, E. Baer, J. Appl. Polym. Sci. 52 (1994) 163–176. (b) S. Nazarenko, D. Hardeski, A. Hiltner, E. Baer, Macromol. Chem. Phys. 196 (1995) 2563–2585.
[30] J.A. Radford, T. Alfrey Jr., W.J. Schrenk, Polym. Eng. Sci. 13 (3) (1973) 216–221.
[31] T. Kazmierczak, H. Song, A. Hiltner, E. Baer, Macromol. Rapid Commun. 28 (2007) 2210–2216.
[32] (a) M.A. Wolak, M. Pan, A. Wan, J.S. Shirk, M. Mackey, A. Hiltner, E. Baer, L. Flandin, Appl. Phys. Lett. 92 (2008) 11301–11304. (b) M. Mackey, A. Hiltner, E. Baer, L. Flandin, M.A. Wolak, J.S. Shirk, J. Phys. D 42 (2009) 1753304.
[33] (a) D. Jarus, A. Hiltner, E. Baer, Polym. Engg. Sci. 41 (12) (2001) 2162–2172. (b) D. Jarus, A. Hiltner, E. Baer, Polymer 43 (2002) 2401–2408.
[34] (a) A.P. Ranade, Structure property relationships in various layered polymeric systems, Chapter 2, Ph.D. Thesis, 2007. (b) M.A. Barger, D. Bland, M.H. Mazor, E. Baer, J. Dooley, J.A. Garcia, WO 2,008,008,875 A2/US7,993,739, 2011. (c) A.P. Ranade, A. Hiltner, E. Baer, D.G. Bland, J. Cell. Plast. 40 (2004) 497–507.

[35] (a) S.R. Armstrong, Novel applications of co-extruded multilayer polymeric films, Ph.D. Thesis, 2013. (b) G.T. Offord, S.R. Armstrong, B.D. Freeman, E. Baer, A. Hiltner, D.R. Paul, Polymer 55 (2014) 1259–1266.
[36] V.A. Topolkaraev, US Patent 7,303,642, 2007.
[37] M. Ponting, T.M. Burt, L.T.J. Korley, J. Andrews, A. Hiltner, E. Baer, Ind. Eng. Chem. Res. 49 (2010) 12111–12118.
[38] J. Du, S.R. Armstrong, E. Baer, Polymer 54 (2013) 5399–5407.
[39] (a) D.L. Krueger, J.F. Dyrud, US Patent 4,729,371, 1988. (b) E. Joseph, J.A. Rustad, US Patent 5,190,812, 1993. (c) E. Joseph, US Patent 5,232,770, 1993. (d) E. Joseph, US Patent 5,258,220, 1993.
[40] J. Wang, D. Langhe, M. Ponting, G. Wnek, L.T.J. Korley, Polymer 55 (2014) 673–685.
[41] (a) M. Gupta, Y. Linn, T. Deans, A. Crosby, E. Baer, A. Hiltner, D.A. Schiraldi, Polymer 50 (2) (2009) 598–604. (b) M. Gupta, Y. Linn, T. Deans, E. Baer, A. Hiltner, D.A. Schiraldi, Maccromolecules 43 (9) (2010) 4230–4239.
[42] A.G. Evans, Z. Suo, R.Z. Wang, I.A. Aksay, M.Y. He, J.W. Hutinson, J. Mater. Res. 9 (2001) 2475–2484.
[43] (a) Y. Wang, H.W. Milliman, J.R. Johnson III, D.M. Connor, N.A. Mehl, D.A. Schiraldi, Polymer 52 (13) (2011) 2939–2946. (b) X. Li, G.B. McKenna, G. Miquelard-Garnier, A. Guinault, C. Sollogoub, G. Regnier, A. Rozanski, Polymer 55 (1) (2013) 248–257.

2 Coextrusion Processing of Multilayered Films

2.1 Introduction

Conventional multilayered films are typically comprised of two to seven polymer materials configured by a 2- to 17-layered feedblock die that continually orders alternating materials into layered domains parallel to the film surfaces. Modifications to conventional, and sometimes unconventional, multilayered coextrusion processes have been demonstrated to increase the achievable number of layers in a coextruded film, sheet, or annular profile from the teens to tens, hundreds, or even thousands of layers. The ability of a polymer processing field to advance coextrusion processing technology for the production of alternatively ordered parallel layers of plastic materials, into alternating micrometer or nanometer thick domains has come over the span of 50+ years of materials, machining, and rheological research and development. An evolutionary review of multilayer to micro- and nanolayer processing techniques has been presented to introduce the current state-of-the-art processing capabilities. Additionally, a discussion of rheological requirements and the effects that result from reducing individual layer thicknesses into micrometer or nanometer scale has been described here.

2.2 Multilayered Film Processing Technologies

Though micro- and nanolayer coextrusion maybe a new topic to some readers, the polymer materials coextrusion process dates back to the early 1920s with the fabrication of rubber tubing for tires [1]. Additional early-patented applications of layered polymer technologies included garden hoses in 1947 and PP/PE water pipes in 1963. These materials were fabricated by combining two conventional polymer-processing extruders together in a two- or three-layered feedblock attached to a flat or annular exit die. An early example of a conventional multilayer feedblock technology enabling the layer coextrusion was patented by DuPont in 1955. Based on the improvements in mechanical strength, fracture toughness, increased barrier to gas transport, and layered film systems became a research topic for the Dow Chemical Company in the 1960s. This was soon followed by the demonstration of the first multilayered

Manufacturing and Novel Applications of Multilayer Polymer Films
http://dx.doi.org/10.1016/B978-0-323-37125-4.00002-2

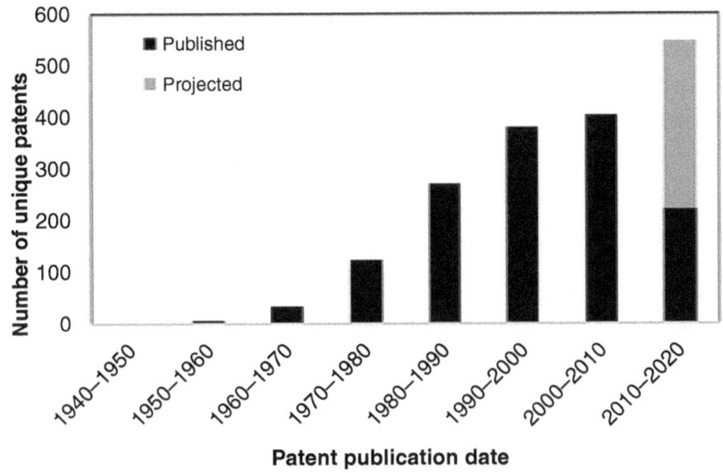

Figure 2.1 Histogram of micro- and nanolayered usage patents published as of June 2015.

films with hundreds of layers and individual layered thicknesses on the order of 100 nm [2]. Over the next five decades, continued research and development in academic and industry arenas enabled continual growth and expansion of the micro- and nanolayered film coextrusion technology to commercial relevancy and products with nearly 500 issued patents for composition of micro- or nanolayered materials applications between the years 2000 and 2010 (Figure 2.1). The trend of increasing research and commercialization of the advanced microlayer technology can be directly attributed to the research and advances in polymer processing as well as the advancement of coextrusion feedblocks and layer multiplying die manufacturing studies and techniques.

2.2.1 Cast Film Microlayer Coextrusion

Conventional 2- to 17-layered cast film coextrusion combines two or more extruders through a multichannel-layered feedblock (Figure 2.2). Separate streams of differing polymer materials are combined into parallel layers in the feedblock before exiting to a film, sheet, or annular die. Multilayered polymer feedblocks, of which some are up to 32 layers, are produced commercially from a variety of polymer die companies including Cloeren, Nordson, Davis Standard, and Macro Engineering. Layered polymer films with less than 20 layers have historically been processed as an improvement over blend films due to performance and cost factors

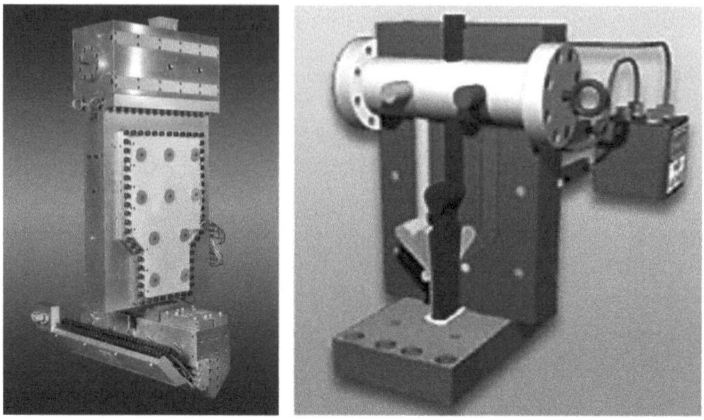

Figure 2.2 Multilayer feedblock designs by Cloeren Incorporated.

that include: (a) potential reduction in expensive polymer material/additive costs through controlling the polymer domain location, continuity, and thickness, (b) incorporate internal film layers of recycle/off-specification materials as filler without degrading film properties, and (c) reduce film thicknesses while maintaining mechanical, transport, and/or optical film properties.

Advances in computer numeric controlled die cutting tools, decades of feedblock and die-design optimization knowledge, and software have enabled drastic reductions in achievable machinability surface tolerances and profiles at the nanometer scale. The ability to increase the layer number, or numbers through variable layer selector plugs in some feedblock designs, in a single feedblock and film die system reached the micro- and nanolayer scale in 2002 [3] with the introduction of Cloeren's NanoLayer™ feedblock (Figure 2.3). Designed and machined to directly produce more than 1000 layers in a single unit, the die connects a designated number of extruders, redistributes the incoming melt streams into hundreds or thousands of layers, which are ordered and distributed within the block using a design inspired from vein splitting. The newly ordered, thousand layer polymer melt stream then exits the feedblock directly into an exit die in order to form the product film or sheet. This commercially available product provides capability and access of layered film tolling manufacturers and converters to process industrial scale quantities of micro- and nanolayer films, thereby providing a ready supply chain for recently developed layered film technologies to enter into the commercial marketplace.

The recent development of layered feedblocks, which is capable of direct coextrusion of hundreds or thousands of layers in cast coextruded

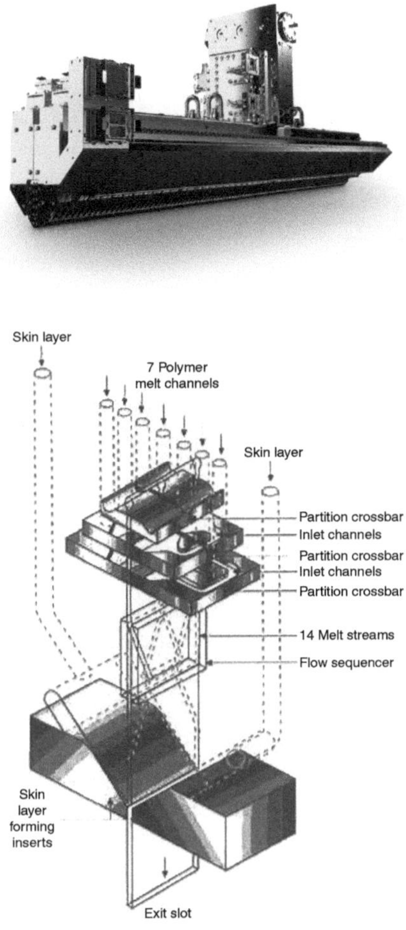

Source: Clorren patent application US 2005/0029691A1

Figure 2.3 Multilayer feedblock designs for nanolayered films by Cloeren Incorporated.

film, is a relatively recent development considering the first examples of micro- and nanolayered technology, which had been demonstrated in the early 1960s. Prior to the fabrication of single shot feedblocks, a combination processing technique utilizing a simple two-, three-, or five-layered feedblock with a series of sequential layer multiplication dies, was initially developed and utilized by the Dow Chemical Company [3,4]. In the sequential layer multiplication die approach of micro- and nanolayering, the two- to five-layered polymer melt stream exits a conventional feedblock and is fed to a series of layer multiplication dies, each of which doubles the

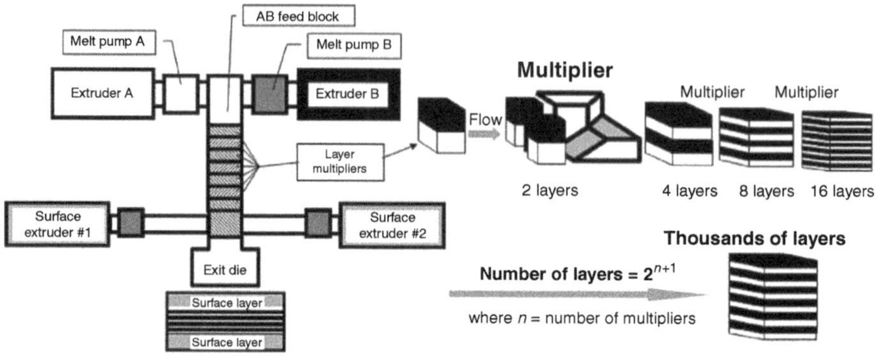

Figure 2.4 (Left) Two component multilayer system comprised of extruders, pumps, feedblock, multiplying dies, surface layer extruders, and exit die. (Right) Schematic illustration of layer multiplication by cutting, spreading, and recombining the melt stream.

number of layers through a process of cutting, spreading, and stacking the layered melt stream (Figure 2.4). A final number of layers in the polymer film is determined as the function of a number of layer multiplication dies, which are placed in series between the feedblock and final film, or sheet exit die. A relationship to the number of film layers can be calculated as a function of: (a) the number of layers in the feedblock and (b) the number of layer multiplying dies. Figure 2.4 depicts one possible case of nanolayer film coextrusion using layer multiplying dies coupled with a two-layered feedblock, which will produce films with a number of layers following a model of 2^{n+1} where n represents the number of sequential layer multiplying dies placed in series between the feedblock and film exit die [5].

The layer multiplication process is very flexible because it allows coextrusion of two or more polymers creating different layer configurations as shown in Figure 1.3. For example, the addition of a third polymer using a three- or five-layered feedblock followed by the layer multiplication process created layered structure with ABC or ATBTA repeating unit. ABC repeating unit represents coextrusion of three different polymers as alternating layers, while ATBTA unit represents a tie layer (T) polymer between alternating layers of polymers A and B. The tie layer polymer is typically used to improve the compatibility or adhesion between the polymers. These structures produced using layer multiplication process can also be combined with skin layers, which are typically added after the layer multiplication dies [5].

In comparison to the single shot feedblock approach for processing micro- and nanolayered films, the layer multiplier die technique is a more flexible and low cost technique. Nanolayered feedblock [6,7] and sequential layer multiplying die [8,9] processing technologies have been demonstrated in this literature to successfully produce nanolayered films with submicron layer thicknesses in a variety of polymer material systems. However, the layer thickness distribution is generally much tighter in a single shot feedblock because sequential multiplication is more susceptible to increased pressure drop issues, variable layer interface contact times, and multiple viscosity based material spreading errors through each die. As a result, the single feedblock processing technique is commonly utilized in production of commercial scale layered film products, whereas the sequential layer multiplying die technology has extensively been used as a research and development tool, through which commercial product formulations and structures have been identified and optimized prior to their commercialization at way lower capital equipment, material, and production costs.

As a result of low cost and increased flexibility of sequential layer multiplying die approach to process layered films having a wide variation in a number of layers, an extensive processing literature exists detailing the capabilities of this layered film processing technique over the past 20 years. Examples of micro- and nanolayered film layer thickness distributions has been published extensively out of Case Western Reserve University by Eric Baer and Anne Hiltner [5,8–10], and more recently at the University of Minnesota [11], and internationally in Universities in Europe [12] and China [13,14]. These laboratory scale units utilize a two or three layer feedblock with a series of layer multiplying dies to produce films with millimeter down to nanometer scale layers in films that are 1–2 in. wide. Examples from Case Western Reserves' microlayer line [5,15], as depicted in Figure 2.5, demonstrate a good layer of parallelism, definition,

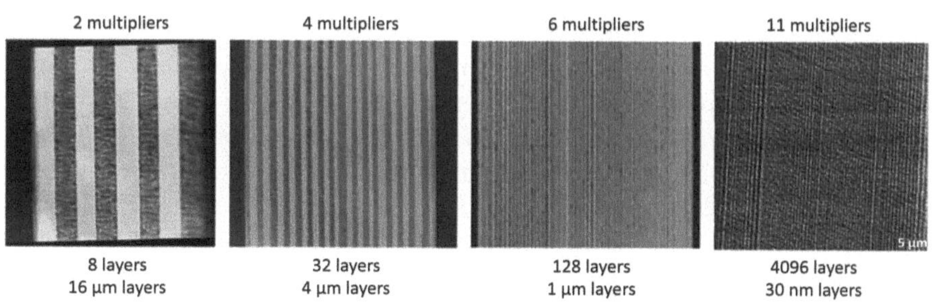

Figure 2.5 Examples of micro- and nanolayers produced from a layer multiplication process with different number of multiplier dies. Multilayered films of PC and PMMA with 50/50 (vol./vol.) composition, the overall film thickness of 127 μm and the number of multipliers used in the coextrusion process with a two-layered feedblock are shown. From left to right, number of multipliers increase from 2, 4, and 6–11. The number of layers and layer thicknesses are as depicted.

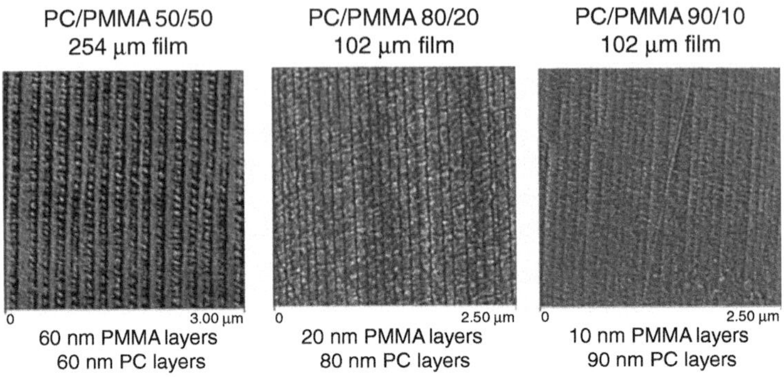

Figure 2.6 Cross-section AFM image. PC (light) and PMMA (dark) layer films with (left) 50/50 composition and 60 nm layers, (center) offset 80/20 composition, and (right) offset 90/10 composition.

and approximately a 10–15% single standard deviation in layer thickness in films with >32 layers.

A previously mentioned, traditional 3- to 17-layered coextruded films provided a benefit of minimizing expensive material costs through enhanced control over the layer material location in a film, surface or internal layers, and relative domain thickness. Micro- and nanolayer coextrusion maintains this advantage through the control of relative compositions/thicknesses of the A and B polymer materials. Coextrusion of variable compositions A/B polymer materials can be accomplished via variation of the extruder system screw speed or melt pump settings to produce different relative A/B material layer thicknesses. An example of variable A/B thickness nanolayered films [5] is shown in Figure 2.6. The polycarbonate (PC) and polymethyl methacrylate (PMMA) layers varied from 1:1 at a 50/50 composition to 9:1 at a relative 90/10 composition. Control of the ratio of A/B layers provides a tool to not only optimize product formulations based on raw material costs, but a means to confine one material in a thinner domain as compared to second layer polymer material. Selective confinement of one polymer will be discussed further in Chapter 3 in relation to enhanced transport barrier in micro- and nanolayered films [5].

Designed mainly as a flexible tool to produce a wide variety of layer thicknesses of nanolayers, significant layer multiplying die design efforts have been published. They highlight the challenges of processing multimaterial, highly ordered polymer blend materials via the coextrusion process. A layer multiplying die design optimization program was reported by the Case Western Reserve University [5] to improve the layer thickness uniformity of sequential layer multiplier dies via modifications

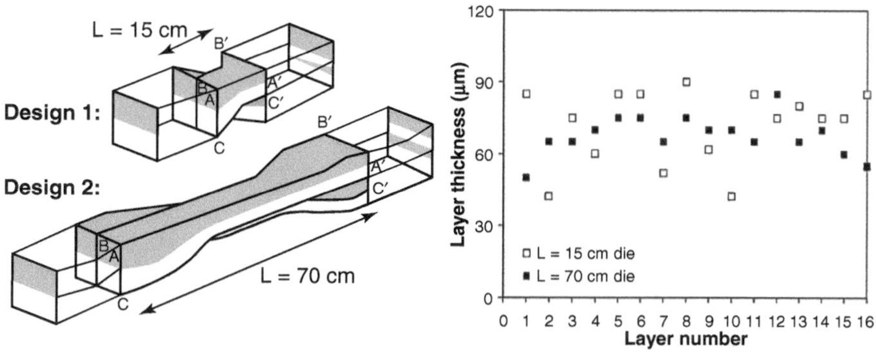

Figure 2.7 (a) Schematic illustrating the difference in polymer flow streamlines based on multiplier die length. (b) the effect of increasing multiplier die length from 15 cm to 70 cm on layer uniformity.

to die path length, which is denoted by *L* in Figure 2.7, in the machine flow direction. Two variations on the layer multiplier die length were fabricated, L_1 at 15 cm and L_2 at 70 cm, and utilized to process a 16-layered PC/PMMA sheet at 50/50 A/B feed ratios. After processing, the sheet samples were cross-sectioned, polished, and imaged under a polarized optical microscopy to measure the layer thickness distribution of every 16-layered film. The longer landing die, L_2, was measured to exhibit a much smaller, 12%, layer thickness distribution (69 ± 8 μm) as compared to the shorter path length die, L_1, which measured a 24% layer thickness distribution (71 ± 17 μm). The improvement in layer thickness deviation was attributed to the longer path length by better equalizing the layered polymer velocity profile during contraction and spreading. These results would suggest that increasing the path length of the layer multiplier dies, or reducing the pitch on the compression/expansion slope inside the die would result in an ideal design. However, a modification of this type would also result in a very long material residence time under elevated processing temperatures in the die. Long polymer residence times under heat are typically nondesirable and avoided as it can result in material degradation and the formation of degradation products such as gels which can disrupt layer uniformity and reduce polymer material properties. This fact, in the context of sequential multiplier die technology requiring multiple dies in series to reach nanolayers, resulted in the Case Western die design study ultimately settling for a 12% thickness distribution, as an achievable measure of the layering precision for this technology [5].

Alteration of the layer multiplying die length is not the only adjustable parameter to alter the layer multiplication performance of the tooling.

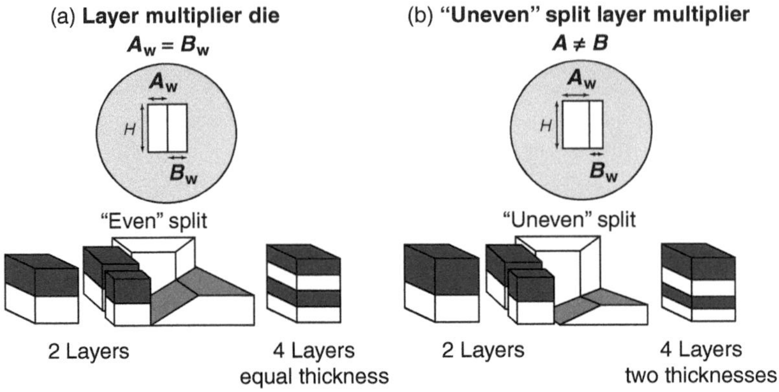

Figure 2.8 (a) Conventional even and (b) uneven split layer multiplier die. The uneven split layer multiplier die produces two layer thicknesses as compared to one with conventional layer multiplication.

Another processing study leveraged the sequential multiplication nature of the die layer multiplication process to capitalize on nonuniform layer thicknesses as a means to build gradient layer thickness distributions in films. Coextrusion through a layer multiplying die redesigned with an unequal vertical split of the entering layered polymer feed stream was demonstrated [16] to produce a layer multiplication from two to four layers, however, the ratio of the layer thicknesses emerging from the die was proportional to the unbalanced A:B split ratio (Figure 2.8) [16].

A mathematical relationship to predict the distribution of layer thicknesses emerging from the unbalanced layer multiplier die was developed as a function equivalent pressure drop in the each of the two die channels, $P_{in} - P_{out}$, as a function of the offset multiplier flow channel width, W_i; height, H; and length, L; layered polymer melt stream viscosity, μ; and layered polymeric volumetric flowrate through the channel, Q_i as shown in Equation (2.1).

$$P_{in} - P_{out} = \frac{12 \times \mu \times L \times Q_i}{H \times W_i^3} \tag{2.1}$$

Setting the pressure drop equal for two unequally split channel dimensions and resolving for ratio of the flowrates, Q_i, enabled the prediction of realized layer split ratios for any geometry of the layer multiplying dies Equation (2.2).

$$\frac{A}{B} = \frac{Q_A / (Q_A + Q_B)}{Q_B / (Q_A + Q_B)} \tag{2.2}$$

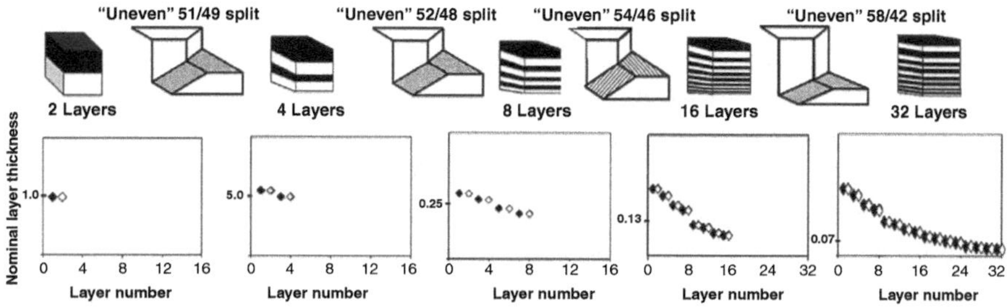

Figure 2.9 Schematic illustration of step-by-step construction of a 10× gradient layer thickness distribution multilayered film by coextrusion through a series of four uneven split layer multiplying dies.

Based on these calculations, a series of four uneven split layer multiplying dies was constructed by Case Western Reserve to achieve unbalanced layer multiplication ratios of: 51/49, 52/48, 54/46, and 58/42. Coextrusion trials were completed utilizing this series of dies to sequentially build a 10× gradient thickness film with layer thicknesses ranging from 30 μm to 3 μm (Figure 2.9) [16].

Good agreement was measured between cross-section optical micrographs of the coextruded film and the layer thickness distribution predictions according to the fabricated layer multiplication ratios predicted from Equation (2.2) (Figure 2.10). Additional flexibility in the fabrication of different shaped layer thickness gradients was demonstrated in coextrusion trials through the unequal layer thickness multipliers, which were

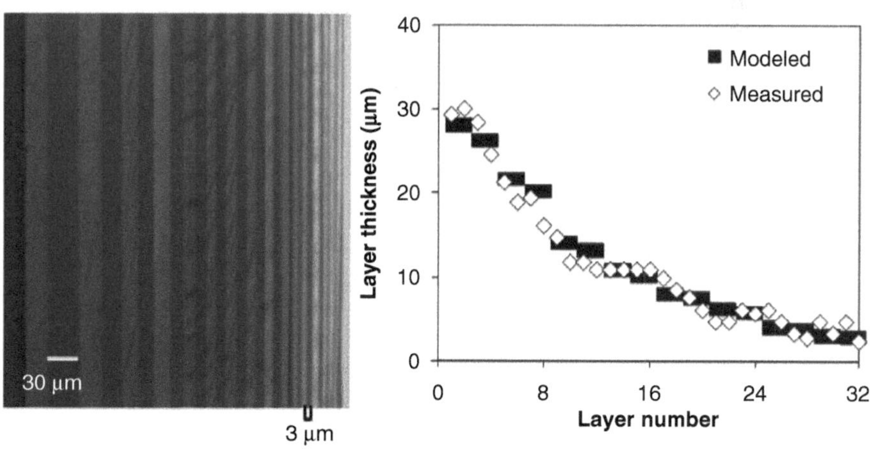

Figure 2.10 Cross-section optical micrograph of a 32-layered PS/PMMA gradient layer thickness film with a 10× difference between the thickest and the thinnest layers.

sequenced in a different order and in combination with traditional evenly split multipliers, in order to demonstrate nanolayer gradient thickness films, as discussed in Chapter 4, for their optical reflective properties [16].

The ability to create numerous gradient layer thickness films and sheets is not solely restricted to the layer multiplying die technology. Fabrication of layered feedblocks, traditional or nanolayered, have been produced to demonstrate films with similar optical properties. The main difference in utilization of a feedblock or layered multiplier die, for the creation of gradient structures, similar to traditional uniform layered films, lies in controlling the uniformity of layer thickness distribution, as well as a flexibility to create several gradient shapes or a number of layers with dies, in contrast to a single layer thickness distribution in a layered feedblock. Again, the selection of a layer multiplier die or a layered feedblock, is not an exclusive option for process development or production. It is the strength of the current state of the micro- and nanolayered film processing technology that both technologies can be utilized to produce film, sheet, or annular structures. The decision for which processing technique is the "best" fit, lies mainly in the requirements of individual technologies, their requirements, and necessary product formulation variability. One additional aspect to consider when discussing or researching layer multiplying dies and layered feedblocks technology is their adaptability to produce nanolayered films through polymer processing techniques outside of cast film, as highlighted in the section on microlayered blown film and other layered processing techniques.

2.2.2 Blown Film Microlayer Coextrusion

Blown film extrusion is extensively utilized to fabricate packaging film, much of which is multilayered to improve mechanical, transport, and thermal properties as required by the food or medical industry. Research efforts, simultaneously conducted at The Dow Chemical Company [17] and Cryovac/Sealed Air Corporation [18], have resulted in process technology advances to enable micro- and nanolayer coextrusion capabilities for blown film processing lines. Following the historical development path of cast coextrusion microlayer film technology, the development of blown film coextrusion multilayered feedblocks with greater than 20 layers, though not yet at thousands, and sequential layer multiplication dies, have been successfully reported. Early versions of blown film technology used spiral mandrel dies where layers are formed by separate spiral manifolds, which are present at different radial distances. The melt from different manifolds, is joined near the exit of the die to create layered structure. However, such die geometry limits the number of layers as the

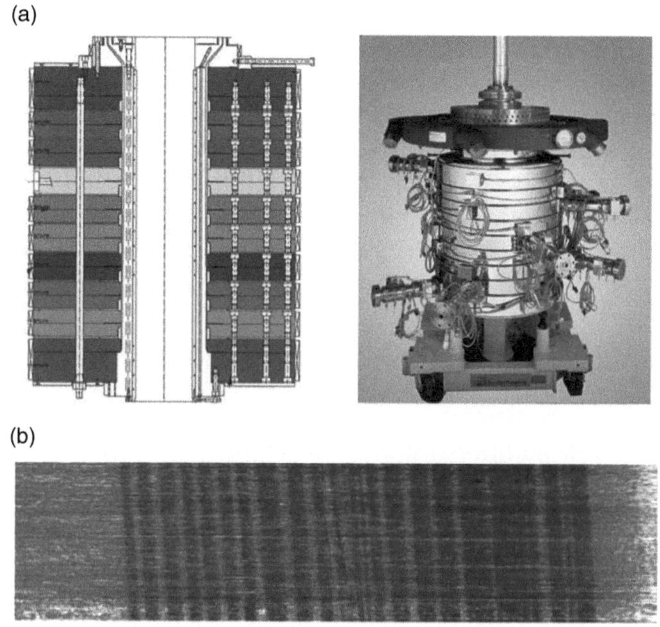

Figure 2.11 (a) Multilayer stacked plate or "pancake" die, and (b) an AFM image showing an expanded view of the core microlayers near the die overlap region in a microlayer blown film containing 30+ layers.

die structure can become significantly large increasing the die complexity, fabrication cost, and material residence time. In another type of die, use of stacked plates with spiral channels on the surface of the plates allowed stacking multiple plates to create layered structures, as depicted in Figure 2.11, which also shows an example of a commercial die with stacked plates. Multilayered blown film lines are currently commercially available from many equipment manufacturers at varying numbers of layers including, but not limited to, Davis Standard, Macro Engineering, Alpha Marathon, Bandera, and Windsor [17].

One of the challenges in adapting the feedblock and layer multiplier die technologies from flat film or sheet to annular structures basically involves ensuring continuous layers around the circumference of the bubble. Nonuniformity or breaks and weld lines caused during the layer wrapping around the circular dies will result in nonuniform film performance and potential premature product failure points acting as high flux pinhole defects in the barrier film. Efforts to combat the continuous wrap issue have resulted in the development of an overlapping section of the circular layered dies, which are demonstrating satisfactory levels of protection against manufacturing defects. An AFM image of the layers near the melt overlap region in the blown film of alternating layers of LDPE and polyolefin plastomer has been shown in Figure 2.11b. Further innovations demonstrated fabrication of 100+ microlayered films in a blown film structure. Achieving uniform wrap of sequentially layered films will continue to challenge the layer multiplier die approach for blown film coextrusion. However, the industry has an advantage of applying over 50 years of polymer processing knowledge to design and investigate the means of producing nanolayered blown film, in one of the fastest growing polymer processing fields.

2.2.3 Other Microlayer Processing Techniques

Besides the commercial cast and blown film coextrusion processing techniques, a few additional multilayer related polymer processing techniques utilize layer multiplying die-like technology to achieve nanometer scale control over multipolymer blend and layered systems. A layer multiplier-like static mixer, which is referred to as the interfacial surface generator (ISG) static mixer [19], is a modular insert to assist in the dispersion of polymer blends or additives. The ISG mixer resembles a four-channel layer multiplication die, that is, two "layers" would multiply to eight, which can be inserted into existing open extrusion transfer pipes or landings. This design results in a lower pressure drop than traditional static mixers, with melt streams multiplied up to 2 million times than that possible for increased blend dispersion uniformity. The utilization of layer multipliers, or similar devices, to control dispersion and domain size of polymer blends has also been demonstrated in packaging films [20, 21].

Chaotic mixing is another polymer melt processing technique, that utilizes a series of layering dies to fold layered polymer melt streams as a path to nanolayered films and sheets. Developed at Clemson University [22–24], the process of chaotic mixing involves passing separate melt streams through a series of sequential shearing stir rods of different geometries to "fold" the materials and melt streams into each other, and create a layered structure (Figure 2.12) [22].

Invoking a shear induced layering mechanism in chaotic mixing toward layering drastically contrasts the flow and material viscosity flow control as employed in machined dies and feedblocks. The chaotic mixing process is a lower pressure process, however it does not achieve uniform, or predictable gradient, layer thickness distributions through the final layered film thickness. However, the technique has successfully demonstrated nanolayer thicknesses in layered articles, in addition to handling mismatched polymer material rheology. The importance of understanding polymer

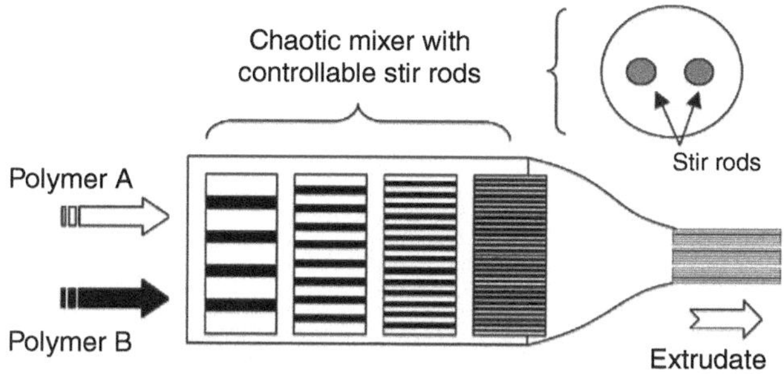

Figure 2.12 Two or more polymer melt streams from separate screw extruders are converted progressively and controllably within a chaotic mixer to multiple layers that may be extruded to give multilayer films or other extrusions.

material rheology is the main driver behind high quality uniform films, during traditional and microlayer coextrusion. Rheological considerations in the selection and processing of micro- and nanolayered films have been discussed in detail in the further section.

2.3 Rheological Phenomena in Multilayered Films

2.3.1 Melt Flow Instability in Coextrusion Process

Melt flow instabilities or "melt disturbances" are inherent to the multilayer coextrusion process due to differences in the non-Newtonian melt flow properties of the component polymers [25–27]. Furthermore, the velocity distribution or the shear rate of the polymer melt flow in the feedblock or the layer multiplier dies is typically parabolic creating flow instabilities. The interfacial distortions can be caused by the flow instability, viscosity, and viscoelastic differences between the component polymers. It is obvious that the interfacial deformation during processing leads to layer thickness nonuniformity, layer thickness variation, irregular interfaces, polymer intermixing, and even film thickness variations.

The flow rate dependent interfacial features are schematically shown in Figure 2.13 [27]. At low melt flow rates, the interfaces are flat or smooth. At slightly increased flow rates, low-amplitude waviness can be observed in the layered structure. Typically, the overall film or sheet quality is not affected with the minor interfacial defects. However, at higher melt flow rates, very strong layer distortions can be noticed in the film structure.

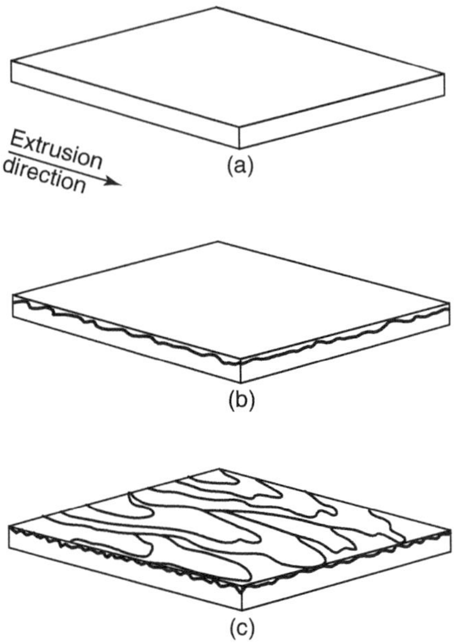

Figure 2.13 Illustration of film or sheet appearance. Under (a) stable flow conditions, (b) incipient interfacial flow instability, and (c) severe instability.

To understand the origins of the layer instability, three-layered sheet of acrylonitrile-butadiene-styrene (ABS) and high impact polystyrene (HIPS) was processed using a single manifold, flat film die and a three layer feedblock [27]. The interfaces along the flow path were investigated under various stable and unstable coextrusion conditions. Two examples of ABS/HIPS/ABS three-layered sheet under stable and unstable processing conditions are shown in Figure 2.14. Layer thickness nonuniformity can be clearly seen in this example. In this study, the important variables affecting the flow instability were identified as skin-layer viscosity, layer thickness ratio, total rate, and die gap. It was concluded that the interfacial flow instability is related to the critical interfacial shear stress, at which the interfacial flow instability occurs for a given multilayer film system. For ABS/HIPS/ABS sheet, it was demonstrated that the interfacial flow instability can be reduced by increasing the skin layer thickness, increasing die gap, reducing coextrusion throughput rate or decreasing the viscosity of the skin layer polymer [26,27].

Other flow instabilities include "zig-zag," "scattering," and "wave patterns" in the extruded sheets [28]. A "zig-zag" pattern is observed as chevrons in the flow directions, which are initiated in the die land above the critical interfacial shear stress. Reducing the shear stress below the critical

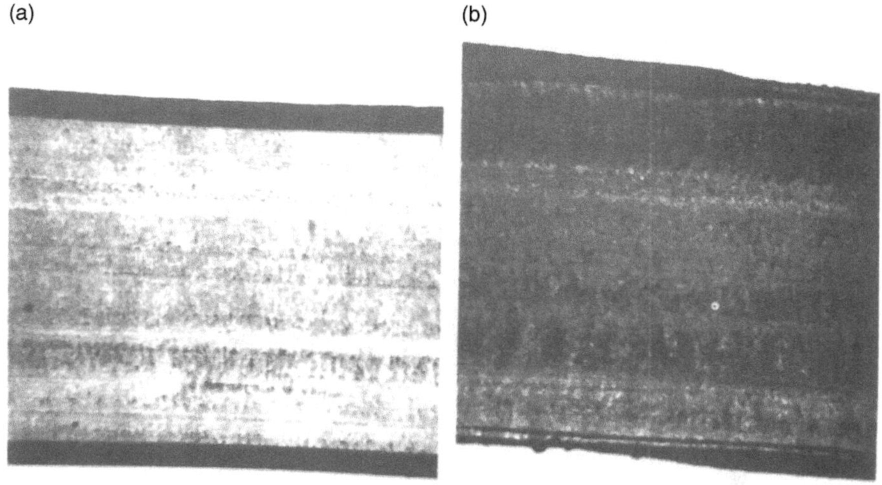

Figure 2.14 Cross-sectional photomicrograph. Through a three-layered ABS/HIPS/ABS sheet made under (a) stable and (b) unstable coextrusion conditions. Thicknesses of ABS and HIPS layers were 0.0013 and 0.0165 in., respectively.

value can help remove this instability. "Scattering" is typically observed in films containing large number of layers (>100), which is also relevant in fabrication of many nanolayered film systems discussed in this book. "Wave" instability appears as multiple parabola shapes spanning the width of the film in the flow direction and is apparent when viewed from the surface. The effect of die geometry, feedblock geometry, and resin rheology were investigated to understand wave instability in multilayer coextrusion. The wave instability originated prior to die land in the die, influenced by the die as well as the feedblock geometry. Different die geometries (and feedblock geometries) with variable expansion ratio, the ratio of the channel width at the exit and the channel width at the die manifold entry, were investigated. It was observed that the increasing expansion ratio and the channel length resulted into increased flow instability for a given polymer system. Furthermore, the structure asymmetry and polymer types were also responsible for flow instabilities. Examples of zig-zag and wave instabilities are shown in Figure 2.15 [29]. The spatial progression of the wave pattern instability showed the wave pattern instability originated at the internal interfaces [26,29].

Viscosity mismatch is one of most common reasons for interfacial distortions. Viscosity mismatch between the two polymers results in migration of low viscosity polymer to the wall, which encapsulates the high viscosity polymers [30,31]. In fact, the degree of encapsulation depends on the degree of viscosity mismatch, shear rates, and the residence time.

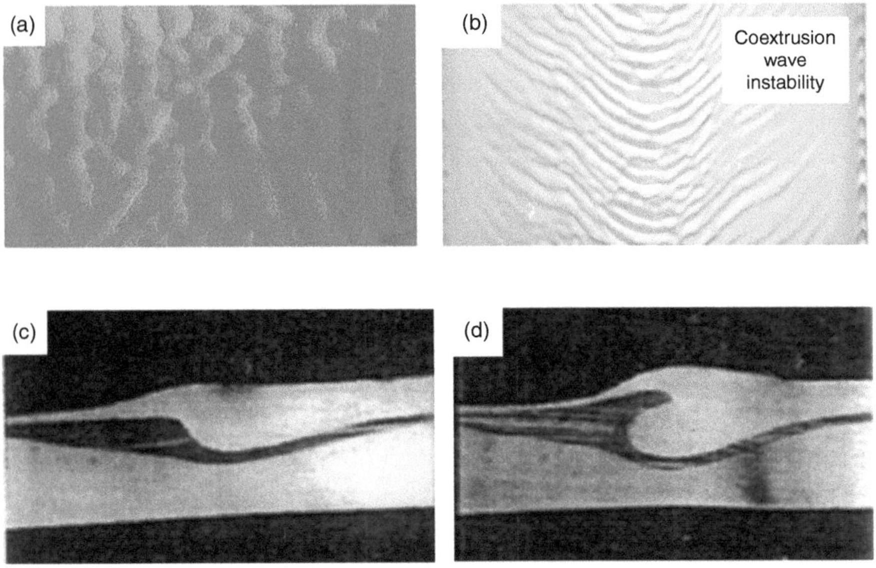

Figure 2.15 Interlayer instability patterns. (a) Zig-zag pattern, (b) wave pattern, and (c,d) cross-section image of wave pattern spatial progression in a three-layered film.

Encapsulation effects in polymer melts with different viscosities have been investigated in tube and slot dies by coextruding low and high viscosity melts together. With sufficiently long contact time, the lower viscosity material encapsulates the high viscosity polymer and low viscosity melt migrates to the wall. Similar observations for flat sheet dies led to increased layer thicknesses of the low viscosity material at the edges creating layer distortions. Differences in the viscoelasticity properties of the component polymers also caused layer distortions. Polymers with relatively high elasticity resulted into layer interface distortion due to the secondary flow patterns normal to the primary flow. The effects are significantly pronounced with increasing flat die width. Furthermore, the multilayered films different elastic properties cause layer interface distortion due to differences in the elasticity between the layers.

These interfacial stabilities are addressed using different methods. For example, increasing the skin layer thickness, increasing the die gap, reducing the material throughput rate, and decreasing the skin layer viscosity are some of the techniques to reduce zig-zag type interfacial instability. Wave instabilities can be minimized by using appropriate feedblock or die design criteria to join the polymer melt streams and minimize the instabilities. Typically, consistent melt flow velocity, parallel layer geometries are preferred to avoid instabilities.

Many early studies focused on understanding the effect of viscosity differences on layered structure [30]. Importance of elasticity ratio in addition to viscosity ratio was also discussed [32]. In recent years, Dooley and coworkers performed a detailed investigation to understand the effect of viscosity and elasticity on the melt flow in the coextrusion process [33]. The results of the viscous encapsulation and the elastic layer rearrangement have been summarized later.

2.3.2 Effect of Viscosity Mismatch on the Layer Structure

Layer uniformity is an important requirement to achieve consistent properties in the final multilayer film product. Among many variable affecting the layer uniformity as discussed earlier, viscosity is one of the most important factor in controlling the layer uniformity. In many cases, the layer distortion in the multilayered film results from significantly different viscosities of the component polymers at coextrusion processing temperatures. As mentioned earlier, viscous encapsulation of the high viscosity polymer by low viscosity polymer occurs as the less viscous polymer moves to the die walls to encapsulate the high viscosity polymer. The effect of the viscosity differences on layer distortions was investigated in two-layered PS/PS structures [26,34]. In this study, four types of polystyrene (PS) resins, Styron 482, 421, 484, and 495 – labeled A, B, C, and D – with different rheological properties were selected as shown in Figure 2.16a. Two-layer structures with 20% skin layer and PS D as a substrate layer were investigated. PS D was selected as a substrate layer as it showed the highest viscosity among these polymers. The viscosity

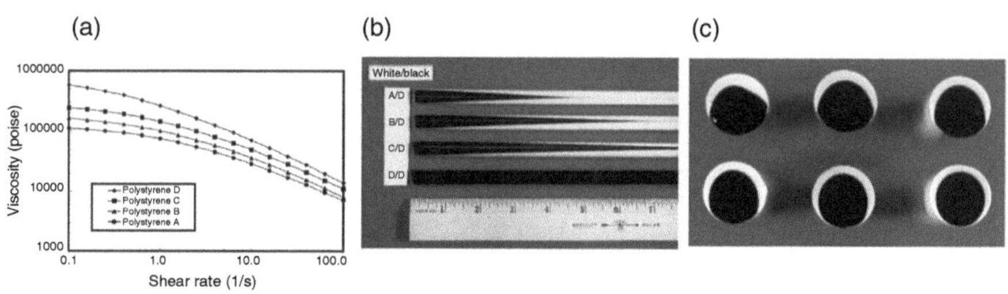

Figure 2.16 (a) Viscosity comparison of the PS resins, (b) samples removed from the circular die channel. The viscosity ratio decreases from 2.5 to 1.0 from samples A/D–D/D, and (c) cross-section images of the A/D samples beginning near the channel entry and then at distances approximately 2.5, 5.2, 7.6, 10.2, and 15.2 cm from the channel entrance.

ratio for A/D, B/D, C/D, and D/D systems were calculated as 2.5, 2, 1.4, and 1, respectively. To investigate the viscous encapsulation effect, a circular channel die was fabricated using two halves, which could be separated to remove the polymer sample. In the coextrusion process, the steady state-conditions were ensured prior to stopping the extruders followed by melt cooling in the die channel. The die channels were subsequently opened to analyze the solidified melt structure. Figure 2.16b showed that the viscous encapsulation in the highest viscosity ratio system, A/D, occurred faster than the other film systems. The encapsulation time increased noticeably with decreasing viscosity ratio. Further analysis of the A/D sample revealed that the encapsulation process initiated at the die channel entry as demonstrated in cross section images of the A/D sample at different die channel locations, Figure 2.14c. The encapsulation velocity was measured to investigate the movement of the cap layer encapsulation of the substrate layer. The highest encapsulation velocity and the shortest distance in the die channel to achieve full encapsulation were observed in A/D system. It is clear that the relative viscosity of the polymer components in the coextruded structure is important to achieve uniform layer structures [26].

Although "viscosity matching" is important in reducing viscous encapsulation effects, layer deformation has been observed in different combinations of polymer resins in coextruded structures. To investigate this effect, two-layered structures with similar viscosities were extruded through circular and square channel dies [26]. Properties of the three polymers – PS, polyethylene (PE), and PC – used in this study are shown in Figure 2.17. Two-layered systems with same polymers as substrate and cap layers and 80/20 compositions were extruded. The polymer melt was cooled in the

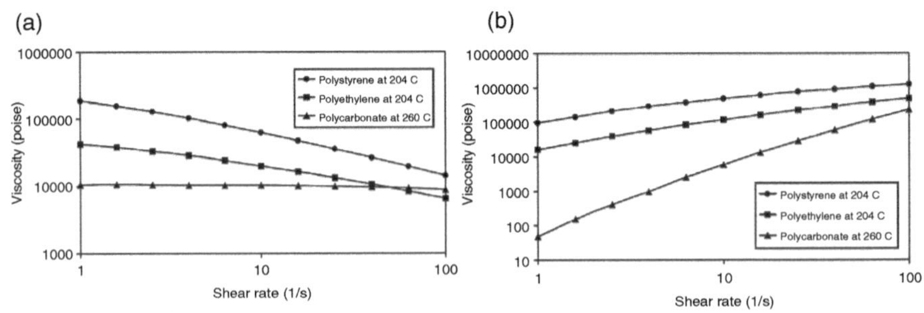

Figure 2.17 PS, PE, and PC resin properties. (a) Viscosity comparison and (b) storage modulus comparison.

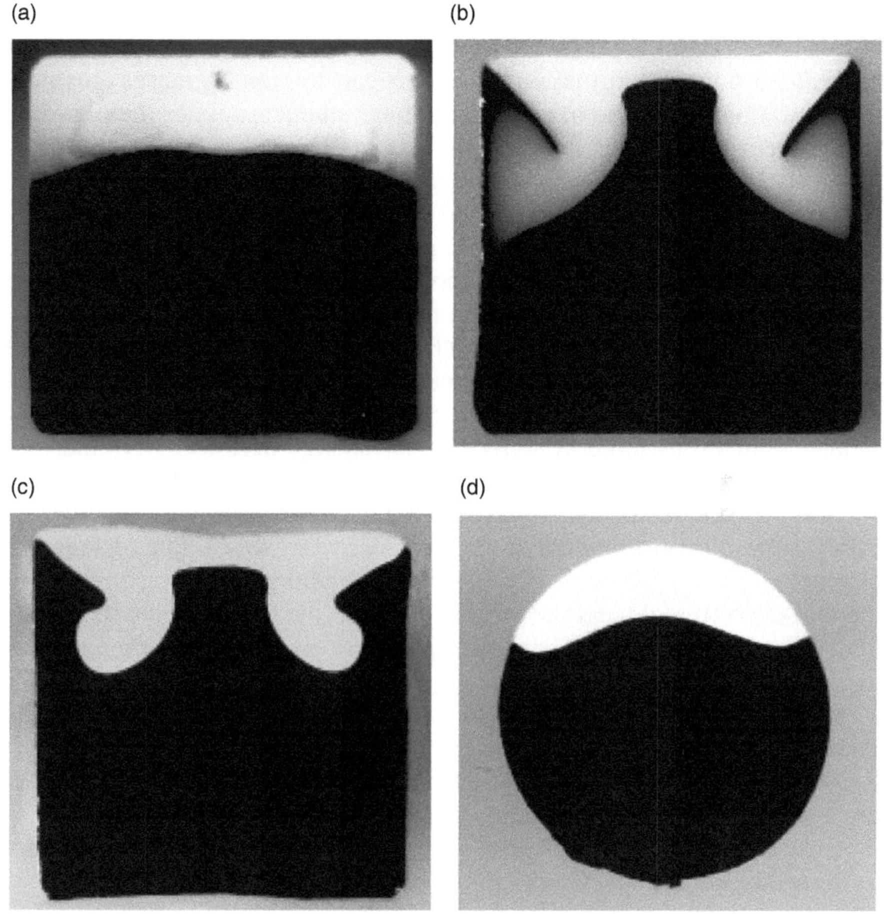

Figure 2.18 Two-layered patterns. Patterns of (a) PC, (b) PS, (c) PE near square channel exit, and (d) PE structure near circular channel exit.

die channel after extrusion steady state was reached and analyzed for layer distortions [26].

The PC/PC system showed fairly flat interfaces, while PS/PS system showed significant distortion in layer interfaces with material flow along the die walls, corners and then toward the center of the channel, Figure 2.18. It is obvious that the viscous encapsulation phenomenon discussed earlier is different than the one observed in similar viscosity materials. It was suggested that the secondary flows in the polymer melt perpendicular to the main flown direction may be present in the polymer melt. The secondary flows are not caused by the viscous effects but produced by the second normal stress differences [31,35].

In the viscous encapsulation process, encapsulation decreases along the melt flow channel as the low viscosity material encapsulates the high viscosity material along the flow length. In contrast, the elastic encapsulation progresses steadily in the melt channel causing melt rearrangement. A similar effect in PE/PE system was observed with less extensive layer arrangement as compared to PS/PS system, possibly due to less elasticity of PE than PS as indicated by the storage modulus plot in Figure 2.17b. The layer rearrangement was least in PC/PC system, which also corresponded to the lower storage modulus than PS and PE. Extrusion of the materials through a circular flow cannel showed significantly reduced layer rearrangement as the secondary flows are absent in a radially symmetric geometry. The layer arrangement in all three systems did not change across the circular die length during coextrusion process. When the circular channel feedblock was used to create concentric layers, no layer distortion was observed in PS/PS system confirming the absence of secondary flows. In a rectangular channel geometry die (4:1), the layer rearrangement was predominant at the corners of the die and the large sections at the center of the die maintained flat interfaces suggesting advantages of high aspect ratio channels in achieving uniform layers [26].

Further investigation on the effect of higher number of layers on the layer arrangement was investigated in 27- and 165-layered feedblock [26]. Examples of 27- and 165-layered PS/PS samples, near the square channel die exit, are shown in Figure 2.19. Similar to a two-layered structure, flow patterns showed significant secondary flows in the polymer melt leading to layer distortions. The results in the rectangular die were also similar to two-layered samples in similar die [26].

To differentiate between the viscous and elastic effect, the layer rearrangement velocities of the layers in the extrudate were measured. The viscous encapsulation dominated initially and reached steady state along the channel length, while the elastic effect remained constant across the channel length and became dominant factor for layer rearrangement. An example of layer movement velocity in sample A/D in a square channel is shown in Figure 2.20a. The results implied that the viscous encapsulation is dominant in a small die, while as the flow path gets longer in a larger dies, the elastic layer rearrangement will dominate the layer distortion. These investigations reveal importance of both viscous as well as elastic components in die designs. The summary of the combination of the flows is shown in Figure 2.20b. A two layer structure with a cap layer (white) less viscous than the substrate layer (black). The arrows on the left indicate the viscous encapsulation flows, while the secondary

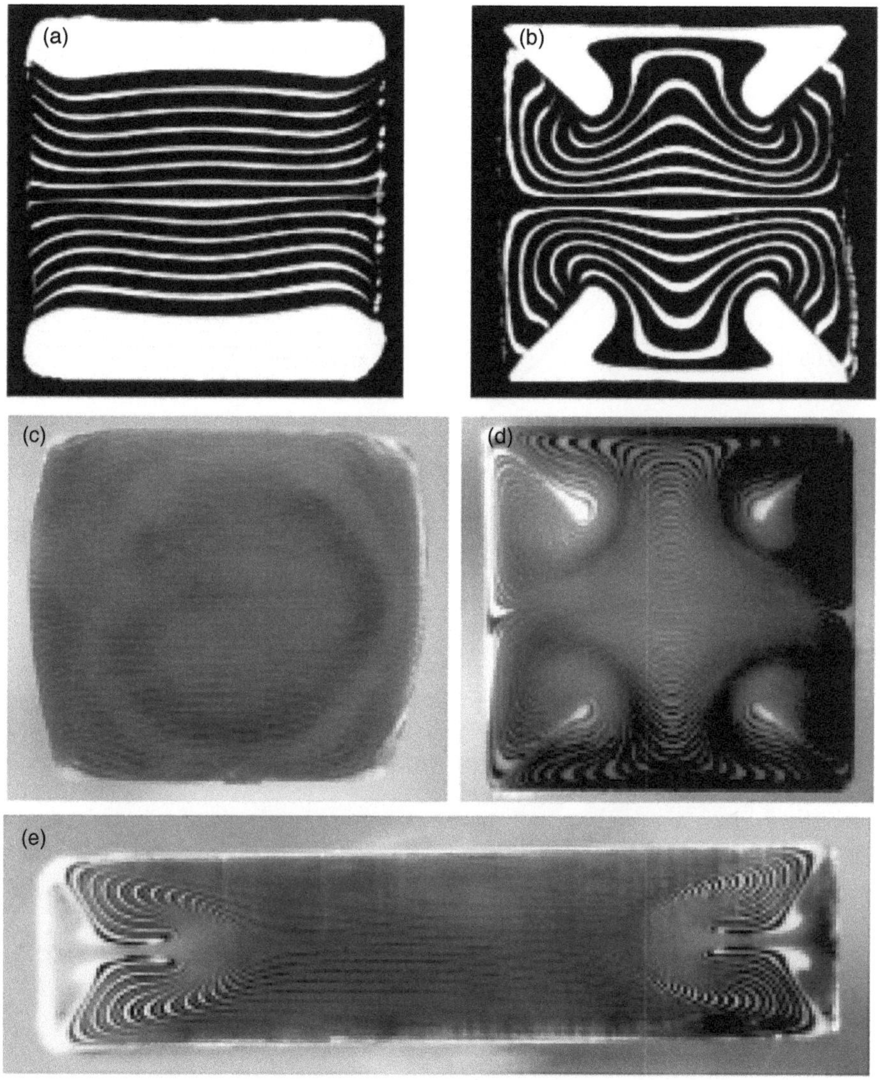

Figure 2.19 (a) A 27-layered PS structure near the entry, (b) the exit of the square channel, (c) a 165-layered PS structure near the entry, (d) the exit of the square channel, and (e) a 165-layered PS structure near the exit of a rectangular channel exit.

flow patterns are shown in black substrate layer. Viscous encapsulation primarily progresses along the die channel wall, while elastic layer rearrangements created flow patterns in different directions. In the coextrusion process, these flow patterns are present simultaneously creating complex flow field [26].

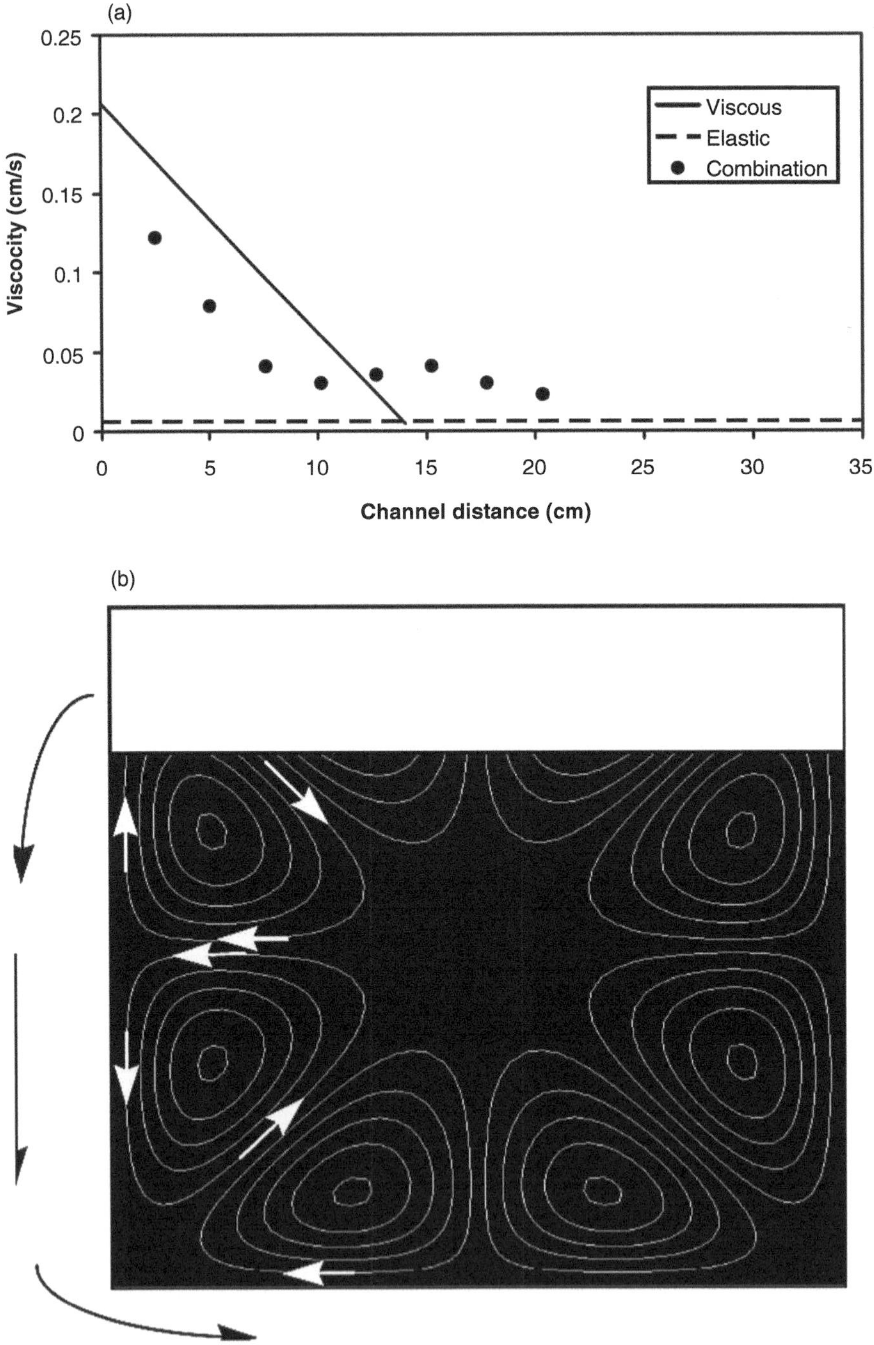

Figure 2.20 (a) Combination of flow patterns produced by viscous encapsulation and elastic secondary flows in a square channel and (b) layer movement velocity for sample A/D in a square channel.

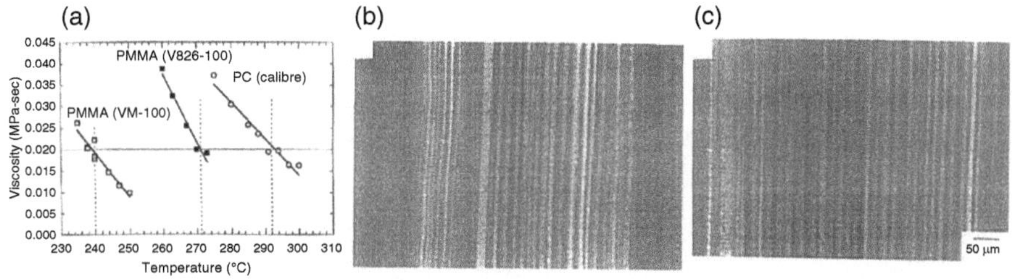

Figure 2.21 (a) PC and PMMA viscosities as a function of temperature with the melt temperature and die temperature used during processing indicated. Optical micrographs of PC/PMMA microlayers have been shown in (b) PC/PMMA1 and (c) PC/PMMA2.

2.3.3 Viscosity Effect in Layer Multiplication Process

The effect of viscosity on a multilayer film coextrusion with different polymers was examined in PC/PMMA multilayered film system [36]. PC (Calibre 200-10) was coextruded with two PMMAs with different molecular weights: (a) PMMA1 = 195,000 g/mol and (b) PMMA2 = 135,000 g/mol. The viscosity date for these resins, measured at a shear rate of 10/s is shown in Figure 2.21a. For the PC/PMMA1 system, the viscosity matched at 292 and 241°C and PC/PMMA2 system viscosity matched at 292 and 271°C, respectively, with a smaller temperature difference for the viscosity match. Sixty-four-layered films with 50/50 composition were produced and the layer structures were compared, as shown in Figure 2.21. In PC/PMMA1 system, poor viscosity match resulted into large layer thickness deviation and only 58 layers were distinguished in structure. In contrast, PC/PMMA2 system with better viscosity match showed all 64 layers and less layer thickness variation. Although the extruder and melt pump temperatures were set at the target values to match the viscosities, the melt temperature equilibrated in the layer multiplier dies, thereby increasing the PC viscosity and decreasing the PMMA viscosities, resulting in the viscosity mismatch. It is obvious that the larger viscosity mismatch in PC/PMMA1 system was significantly larger than PC/PMMA2 system causing layer encapsulation, as discussed earlier. It must be noted that the secondary flow effects may have also contributed to layer thickness deviation in both systems.

Coextrusion processing of elastomeric materials is still in research phase. Recently, the coextrusion processing of unvulcanized rubber and elastomers was attempted [37]. Although, multilayered samples with matched and mismatched rheological properties showed layered structure, a significant deviation in layer thicknesses was observed. A significantly

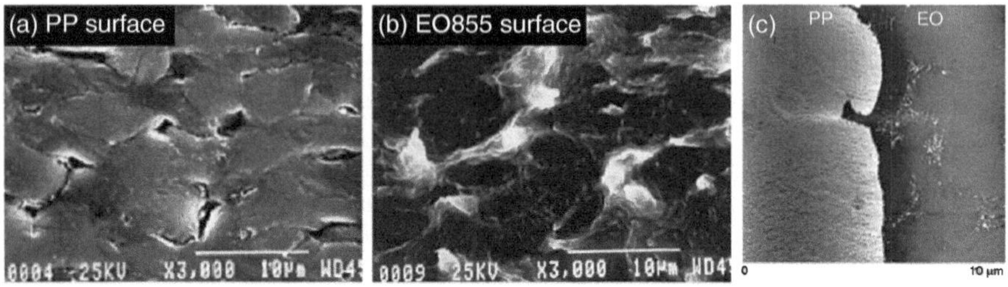

Figure 2.22 Example of peeled surface from PP/EO multilayer system. (a) PP surface, (b) EO surface, and (c) AFM cross-section image of the PP/EO interface with an influx.

large layer thickness coefficient of variation of at least 44% was observed in a 32-layered polyisoprene/butyl rubber system. Continued research and development of new layer multiplication feedblocks and dies, is desired in this field [36].

2.3.4 Melt Influxes and Layer Interfaces

In addition to the viscoelastic properties, crystallization properties also play an important role in the interfacial distortion at microscale. The multilayered films with crystallizable polymers, showed the formation of crystalline influxes between incompatible interfaces of semicrystalline and elastomeric polymers. For example, the volume contraction during PP crystallization created a vacuum trapping of PE melt before its solidification and large number of influxes resulted in modulated surface morphology in the layered structure [38]. Multilayered tapes of PP with ethylene–octene (EO) copolymer and olefinic block copolymers (OBC) showed the presence of influxes as shown in Figure 2.22. Furthermore, the PP/elastomer layered composites also showed undulating periodicity, which was attributed to the PP–spherulite boundaries. In crystallizable polymer systems, the thermal history (melt cooling rate, crystallization conditions, and annealing steps) is critical in the formation of melt influxes or spherulites affecting the layer boundaries. If the layered samples with crystallizable polymers are quenched fast enough during film take-off, it is possible to reduce melt influxes and spherulite size [38].

2.3.5 Interfacial Slip in Nanolayers

Processing either blown or cast layered films, from a multilayered feedblock, or a series of sequential multiplying dies, can result in a significant increase in the contribution of the overall film structure and property from layer-to-layer boundary material mixing regions, which are also called layer interphase regions or volume, as the number of layers is increased to fifty or hundreds in a single film. Directly proportional to the polymer material chemical solubility or Flory–Huggins interaction parameter [39], the variable thickness of this layer interphase is comprised of interwoven polymer chains of every single adjacent polymer layer (Figure 2.23a). Depending on the chemical attraction or repulsion, this layer interface blend region may be of a lower density than the constituent bulk layered materials. Especially

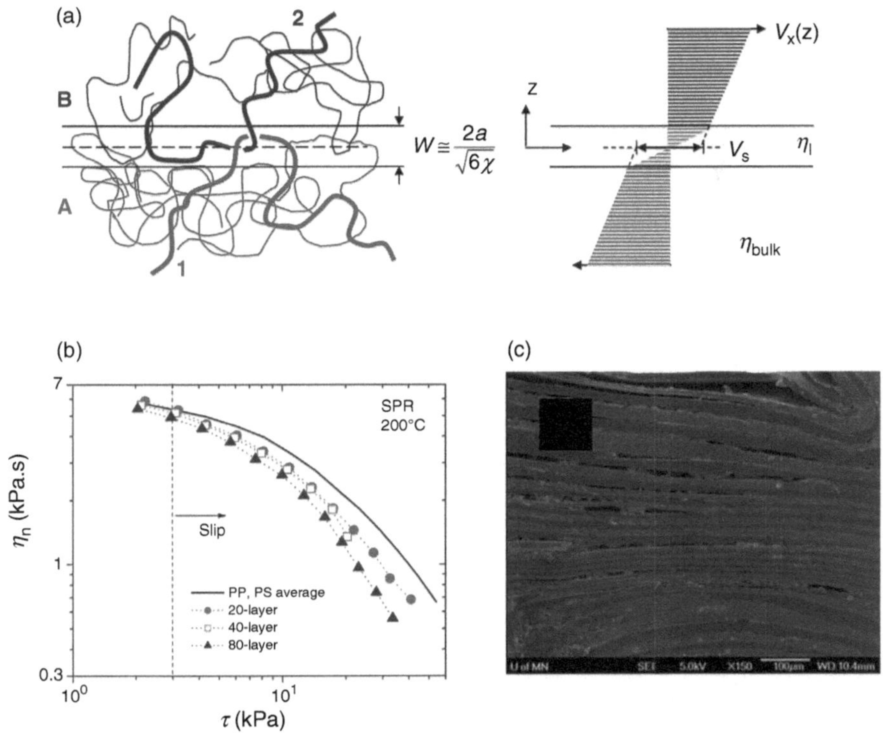

Figure 2.23 (a) The interface between entangled melts of two incompatible polymers A and B. Chains are less entangled in the interfacial region than in the bulk, although some pairs of A and B chains (such as, the chains 1 and 2 shown) nonetheless form entanglements across the interface. The result is a lower interfacial viscosity, η_I, when a shear stress, τ, is applied, the fluid velocity appears, at a macroscopic level, to be discontinuous across the interfacial region, with an apparent slip velocity V_{slip}; (b) nominal viscosity of multilayer samples and the harmonic average solid lines of the neat components, as measured in the sliding plate rheometer by steady shear experiments namely PP and PS, deviations from the average viscosity were observed above approximately 3 kPa; and (c) SEM micrographs of 80-layered PP/PS samples after steady simple shear in the rotational parallel-disk.

apparent in the cases where the A/B layered materials do not have high chemical compatibility or adhesion, densitometry variations within layered films with high numbers of layer interfaces can result in nontrivial layer interphase volumes. Layer interphase regions with lower density, as a result of the reduced polymer chain entanglement, have been shown to induce layer interface instability, when processed films are exposed to high shear rates through a mechanism attributed to interfacial slip [40].

Macosko and coworkers [41,42] have extensively studied the phenomenon of interfacial slip in various layered polymeric film systems, as a function of the number of layer interfaces, material compatibility, and applied shear. Significant film layer instability, characterized by differences in the measured layered melt viscosity (Figure 2.23b), layer break-up, and/or delamination, was shown in layered film systems with >80 individual layers and poor constituent layered material adhesion/chemical compatibility. A cross-section SEM micrograph of an 80-layered PP/PS film demonstrates the breakup of layers and nonuniformities, which can result due to the interfacial slip phenomenon (Figure 2.22c). Consideration should be given to shear induced layer nonuniformity, after layer multiplication has been achieved during melt processing, as most commercial multilayered and nanolayered structures, whether blown or cast, are coextruded into its final form through a high aspect ratio expansion in a film or sheet exit die. The coextrusion film or sheet die casting or bubble expansion process exemplifies the exact type of shear conditions, under which interfacial slip can occur. During film forming extrusion processes, commonly the layered melt stream thickness gets reduced by 50×, while simultaneously spreading the melt stream laterally by a factor of 100× or more, to induce high shear history prior to solidify the layered film product. An interfacial slip phenomenon illustrates the importance of understanding material interactions, and formulating adequate adhesion between micro- and nanolayered materials during processing, beyond the common requirement of viscosity matching materials. Eliminating the possibility of interfacial slip, or processing nonuniformities, can be critical if highly regular, parallel layered interfaces, and thicknesses are required in the material target applications, especially in the case of optical applications of micro- and nanolayered film systems discussed in Chapter 4 [41].

2.4 Summary

The frequency of published and patented micro- and nanolayered film system literature has steadily increased in 50 years, since layer multiplier dies first demonstrated an ability to control layered polymer materials down to the nanoscale. A strong foundational technology base to process nanolayered films has been simultaneously developed in academic and commercial arenas. Process technology equipment, that enables rapid research and development investigation of nanolayer film formulation, via sequentially layered multiplying dies, and commercial scale production of highly uniform, industrial scale throughputs, via nanolayered feedblocks, are currently available for cast film coextrusion lines. The translation of micro- and nanolayered technologies into modern consumer goods like shoes, food packaging, and optical reflectors for architectural applications, highlights the power of having a complete technology "research to manufacturing" capable supply–chain. Success in producing nanolayered cast film products has initiated research efforts to translate nanolayer-like control of polymer layered film processing to blown film processing, for a wider spread product development. Studies about measuring and considering the rheological effects of materials were discussed as a critical process variable toward the successful coextrusion of nanolayered polymer films. Though nanolayer coextrusion can impart additional rheological based processing concerns such as interfacial slip, melt influxes, and magnified layer thickness variation, the first order concern of traditional three-, five-, or seven-layered coextrusion to best matching material in process shear rate viscosities still holds. The increased understanding, measuring, and modeling of rheological properties of polymer materials as a result of nanolayer processing, should only move to inspire the next generation of layered feedblocks and layer multiplying dies that are capable of producing better layer uniformity, limiting pressure drop, or expanding acceptable material viscosity mismatch ratios for layered processing.

References

[1] D. Djordjevic, in: S. Ward (Ed.), Coextrusion, iSmithers Rapra Publishing, Shawsbury, United Kingdom, 62 (1992), pp. 3–4.
[2] The Dow Chemical Company, US Patent 3,195,865, 1965.
[3] Peter F. Cloeren, US Patent Application US 2005/0,029,691 A1, 2005.
[4] The Dow Chemical Company, US Patent 3,759,647, 1973.
[5] M. Ponting, A. Hiltner, E. Baer, Macromol. Symp. 294 (1) (2010) 19–32.
[6] 3M Innovative Properties, US Patent 6,111,696, 2000.
[7] M.F. Weber, C.A. Stover, L.R. Gilbert, T.J. Nevitt, A.J. Ouderkirk, Science 287 (5462) (2000) 2451–2456.
[8] C. Mueller, J. Kerns, T. Ebeling, S. Nazarenko, A. Hiltner, E. Baer, Polym. Process Eng. 97 (1997) 137–157.
[9] R.Y.F. Liu, A.P. Ranade, H.P. Wang, T.E. Bernal-Lara, A. Hiltner, E. Baer, Marcomolecules 38 (2005) 10721–10727.

[10] C.D. Mueller, S. Nazarenko, T. Ebeling, T.L. Schuman, A. Hiltner, E. Baer, Polym. Eng. Sci. 37 (2) (1997) 355–362.
[11] P.J. Cole, C.W. Macasko, J. Plast. Film Sheet. 16 (3) (2000) 213–222.
[12] N. Alipour, U.W. Gedde, M.S. Hedenquist, S. Yu, S. Roth, K. Bruning, A. Vieyies, K. Schneider, Eur. Polym. J. 64 (2015) 36–51.
[13] L. Ren, J. Zhao, S.J. Wang, J.W. Zha, G.H. Hu, Z.M. Dang, Compos. Sci. Technol. 107 (11) (2015) 107–112.
[14] S. Yeng, H. Yu, F. Lei, J. Li, S. Guo, H. Wu, J. Shen, Y. Xiong, R. Chen, Macromolecules 48 (12) (2015) 3965–3973.
[15] T. Kazmierczak, H. Song, A. Hiltner, E. Baer, Macromol. Rapid Comm. 28 (2007) 2210–2216.
[16] M. Ponting, T.M. Burt, T.J. Korley, J. Andrews, A. Hiltner, E. Baer, Ind. Eng. Chem. Res. 49 (23) (2010) 12111–12118.
[17] J. Dooley, J. Robacki, S. Jenkins, P.C. Lee, R. Wrisley, Proceedings of ANTEC 2011 SPE (2011).
[18] Cryovac Inc., US Patent 8,876,512, 2014.
[19] Charles Ross and Son Company, Mixing technology insight #85, accessed online http://www.mixers.com/insights/mti_85.pdf, 2015.
[20] G. Zhang, X. Hu, K. MacInnis, E. Baer, Polymer 57 (28) (2015) 117–124.
[21] H.J. O'Donnell, D.C. Peck, P.L. Caruso, US Patent Application US2012/0,237,746 A1, 2012.
[22] D.A. Zumbrunnen, S. Inamdar, Chem. Eng. Sci. 56 (2001) 3893–3897.
[23] O. Kwon, D.A. Zumbrunnen, J. Appl. Polym. Sci. 82 (2001) 1569–1579.
[24] D.A. Zumbrunnen, S. Inamdar, O. Kwon, P. Verma, Nano Lett. 2 (10) (2002) 1143–1148.
[25] E. Mount III, in: J.R. Wagner Jr. (Ed.), Multilayer Flexible Packaging, Elsevier, 2010, pp. 75–96 (Chapter 6).
[26] J. Dooley, Viscoelastic flow effects in multilayer polymer coextrusion, Ph.D. Thesis, 2002.
[27] W.J. Schrenk, T. Alfrey Jr., in: D. Paul, S. Newman (Eds.), Polymer Blends, vol. 2, Academic Press, 1978, pp. 129–164 (Chapter 15).
[28] (a) W.J. Schrenk, N.L. Bradley, T. Alfrey Jr., H. Maack, Polym. Eng. Sci. 18 (1978) 620. (b) C.D. Han, R. Sherry, Polym. End. Sci. 18 (1978) 180. (c) R. Ramanathan, W.J. Schrenk, I. Whearley, US Patent 5,269,995, 1993.
[29] R. Ramanathan, R. Shanker, T. Rehg, S. Jons, D.L. Headley, W.J. Schrenk, SPE ANTEC Proc (1996) 224–228.
[30] (a) J.H. Southern, R.L. Ballman, J. Appl. Polym. Sci. 20 (1993) 175. (b) C.D. Han, Multiphase flow in polymer processing, Academic Press, New York, (1981) pp 387–418. (c) A. Khan, C.D. Han, Trans. Soc. Rheol. 20 (4) (1976) 595–621.
[31] (a) J.L. White, R.C. Ufford, K.R. Dharod, R.L. Price, J. Appl. Polym. Sci. 16 (1972) 1313. (b) B.L. Lee, J.L. White, Trans. Soc. Rheol. 18 (1974) 467.
[32] M.A.M. Arvedson, J. Plast. Film Sheeting 1 (1985) 22–29.
[33] (a) J. Dooley, B.T. Hilton, Plast. Eng. 50 (2) (1994) 25. (b) J. Dooley, R. Ramanathan, SPE ANTEC 40 (1994) 89. (c) J. Dooley, K. Hughes, Tappi J. 79 (4) (1994) 235. (d) J. Dooley, L. Diestche, Plast. Eng. 52 (4) (1996) 37.

(e) J. Dooley, K.R. Hughes, SPE ANTEC 42 (1996) 236. (f) Anderson, P.D., J. Dooley, H.E.M. Meijer, Appl. Rheol. 16 (4) (2006) 198–205.

[34] J.L. Zyrd, J. Dooley, SPE ANTEC 1 (1998) 217–221.

[35] (a) J. Dooley, L. Dietsche, Plast. Eng. 52 (4) (1996) 37. (b) B. Debbaut, T. Avalosse, J. Dooley, K. Hughes, J. Non-Newt. Fluid Mech. 69 (2) (1997) 255. (c) B. Debbaut, J. Dooley, J. Rheol. 43 (6) (1999) 1525.

[36] (a) C. Mueller, J. Kerns, T. Ebeling, S. Nazarenko, A. Hiltner, E. Baer, Polymer Process Engineering 97, Cambridge University Press, United Kingdom (1997) 137–157. (b) E. Baer, J. Kerns, A. Hiltner, in: A.M. Cunha, S. Fakirov (Eds.), Structure Development During Polymer Processing, Kluwer Academic Publishers, Netherlands, 2000, pp. 327–344.

[37] P.J. Harris, S.O. Carson, J.L. Gadley, J. Maia, Polym. Eng. Sci. 55 (7) (2015) 1520–1527.

[38] P. Dias, Y.J. Lin, B. Poon, H.Y. Chen, A. Hiltner, E. Baer, Polymer 49 (2008) 2937–2946.

[39] R.Y.F. Liu, Y. Jin, A. Hiltner, E. Baer, Macromol. Rapid Comm. 24 (16) (2003) 943–948.

[40] R. Zhao, C.W. Macosko, J. Rheol. 46 (1) (2002) 145–167.

[41] P. Lee, H.E. Park, D.C. Morse, C.W. Macosko, J. Rheol. 53 (4) (2009) 893–915.

[42] J. Zheng, T.P. Lodge, C.W. Macosko, J. Rheol. 50 (1) (2006) 41–57.

3 Gas Transport, Mechanical, Interphase, and Interdiffusion Properties in Coextruded-Multilayered Films

3.1 Introduction

Multilayer coextrusion processing allows the combination of two or more different polymers with unique properties in composite and film structures with individual layer thicknesses ranging from a few tens of micrometers to a few nanometers. With advancements in the new extrusion feedblocks and layer multiplying technologies, production of the composites with nanoscale structures using commercial polymers has enabled the exploration of application relevant properties. Polymer processing parameters, coextrusion techniques, and factors affecting the coextruded structures, have been discussed in detail in Chapter 2. In this chapter gas transport and mechanical properties of the multilayered films, and composites have been reviewed. Other phenomena of interphase and interdiffusion in nanolayer films are also included for their relevance in controlling film properties.

Recent advancements in the area of confinement of crystallizable polymers at nanoscale in multilayered films, showed the creation of highly oriented lamellar morphology impacting the gas transport properties by one to two orders of magnitude. The improvement in the gas barrier properties via multilayer coextrusion at nanoscale offers a unique opportunity to go beyond conventional approaches of barrier improvement as well as allows the usage of commercial polymers for high value-added packaging applications. Many examples of multilayered structures with hundreds of layers and various parameters impacting the gas permeability have been demonstrated.

Layered structures exhibited unique improvements in the adhesion, mechanical, and failure properties. Combinations of a wide range of polymers were investigated to understand the structure–property relationships at micro- and nanoscale. Multicomponent heterogeneous polymers with controlled layer thicknesses allowed systematic development of composite materials with synergistic effect of individual components. Use of compatibilizers as a tie layer, was also explored as it allowed the systematic variation of layer thickness down to a few tens of nanometers, to investigate

Manufacturing and Novel Applications of Multilayer Polymer Films
http://dx.doi.org/10.1016/B978-0-323-37125-4.00003-4

adhesive properties. Recent studies using gradient film structures and block copolymers (BCPs) have been discussed.

As multilayer films create hundreds to thousands of interfaces in the composite, it is intuitive that the interfacial structure plays an important role in controlling the composite properties. Interface phenomena such as interphases in immiscible polymer systems and interdiffusion in miscible polymers are particularly interesting for their ability to impact thermal, gas transport, mechanical, optical, and many other properties. When two components of the coextruded-multilayered structure consist of immiscible polymers, interfacial interaction resulted into interphase materials. Such interphase materials are formed by localized mixing of polymer chains during extrusion. Coextrusion processing of layers down to only a few nanometers produced multilayered films with large amounts of interphase materials, which are usually difficult to achieve by conventional blending techniques. When the two polymers are miscible in nature, controlled interdiffusion in coextruded structures under thermal annealing allowed the investigation of diffusion parameters in the polymers. Examples of multilayered films discussing interphases, interdiffusion, and their relevance in polymer composites, have also been discussed in this chapter.

Review of different properties and structure–property relationships will facilitate the use of these multilayered composites in commercial applications, beyond conventional layered structures and also provide guidance in material selection.

3.2 Gas Transport Properties of Multilayered Films

Polymer materials are widely used in packaging applications for a controlled atmosphere. For example, oxygen permeability control with high volatile organic barrier requirement is required for improving the shelf life of many food and beverages such as milk, alcohol, baby food, etc. [1]. High barrier films are desired in many healthcare and pharmaceutical applications such as blister packaging and IV bags. Conventional approaches to develop high-value-added barrier materials include reactively treated laminates, use of barrier coatings, inorganic fillers, and blends among many others [1]. Multilayered films with only a few layers with barrier polymers have also been developed for such applications. Polymer blends have widely been studied as an economical approach in the development of barrier composites. Many of these techniques are inefficient in producing high barrier films or use expensive polymers in composite structure. Multilayer coextrusion approach is particularly interesting in achieving nanoscale confinement unlike conventional techniques, which use rate-limiting lab scale process such as spin-coating, self-assembly in custom synthesized BCPs and are difficult to scale up. Recent innovations using coextrusion processing allows production of multilayered films with thousands of layers has enabled fabrication of nanoconfined layers.

Gas permeation through polymer films is governed by the properties of polymer, type of permeant molecules, and interaction between polymer and gas molecules. Properties of the polymers greatly vary based on various properties such as chemical structure, crystallinity, processing techniques, etc. In commercially produced thermoplastic materials, the crystallization behavior impacts gas permeability properties, as the chain packing significantly affects the free volume available for gas transportation. The crystalline structures are typically considered impermeable to the diffusing permeant, while gas transport mainly occurs through the amorphous phase in crystalline polymers. Amount of crystallinity and lamellar orientation can be controlled by varying processing conditions. Permeability coefficient (P) of the film is a product of the diffusion coefficient (D) and solubility coefficient (S), given as $P = D \times S$. A basic two-phase model consisting of impermeable crystalline phase dispersed in permeable amorphous matrix can be applied to determine gas transport properties of polymer films. The modifications to the two-phase models have been suggested to account for variable amorphous density, and phase dedensification of the polymer materials. Many studies have investigated the effect of solid-state structure on the gas transport properties of polymers and products. Controlling the morphological properties such as crystal orientation and free volume, have been used to control the gas transport properties.

The lamellar orientation control in crystalline polymers using coextrusion processing approach has led to the development of new confined and highly oriented structures at nanoscale using commercial polymers. New morphologies in the multilayered films showed dramatic impact on gas transport and water vapor transport properties. Various multilayered films contain hundreds of layers with individual layer thickness as low as 20 nm scale has resulted in the formation of highly oriented lamellar structures.

In this section, the structure of polymers in multilayered structure at micro- and nanoscale and its impact on gas transport properties, has been reviewed. Multilayer coextrusion provides an opportunity to achieve one-dimensional (1D) confinement of the polymers by varying layer thicknesses from a few hundred micrometers to tens of nanometers. Relative polymer compositions, film thicknesses, and number of layers can easily be changed to achieve thicknesses ranging from few hundred micrometers down to a few tens of nanometers.

Multilayered films with a composite structure of $[A/B/A]_n$, as discussed in Chapter 2, were produced with a varying number of layers and compositions. Effect of confinement of polymer B, defined as "confined polymer" by polymer A, defined as "confining polymer" was investigated. Numerous combinations of varying sets of commercial polymers have demonstrated improved barrier properties. The confined polymers reviewed in this chapter are: polyethylene oxide, polycaprolactone, syndiotactic polypropylene (sPP), high-density polyethylene, polyvinylidine fluoride homopolymer and copolymer, and poly-4-methyl-1-pentene. Other composite systems with confined polymers as core components in the biaxially stretched films have also been discussed for use in commercial packaging applications.

3.2.1 Polyethylene Oxide (PEO)

PEO was investigated in nanolayer confinement with different confining substrate such as amorphous polystyrene (PS) and ethylene acrylic acid (EAA) copolymer. Multilayered films of PEO/PS with PEO layer thickness ranging from 35 μm to 21 nm were produced by varying number of layers from 9 to 1025. The oxygen permeability of the films with equal volume fractions of PEO and PS layers showed more than two orders of magnitude decrease with decreasing layer thickness, as is shown in Figure 3.1a [2,3]. Using PEO and PS permeability values of 0.38 and 2.08 barrer, respectively, a simple series model as shown in Equation (3.1) predicts the film permeability for PEO/PS (50/50) composite as 0.64 barrer. A nine-layered film showed oxygen transport properties close to the estimated values. However, as the number of layers increased, the oxygen permeability decreased significantly, close to two orders of magnitude. Because PS is an amorphous polymer and the PEO has significantly lower permeability than the PS, the permeability changes in the multilayered films were attributed to permeability changes in PEO layers. It is also obvious from Equation (3.1) that the film permeability will be significantly impacted by low permeability polymer PEO rather than PS permeability, even at low amounts of PEO in the layered structure. Assuming the PS permeability constant, PEO oxygen permeability was calculated and plotted as shown in Figure 3.1b. Another multilayer film system with PEO confinement by low crystallinity polymer, EAA, also showed a similar decrease in the permeability values. The differences in permeability values of PEO were attributed to differences in its structure in the two systems as discussed further.

$$P_{\text{Film}} = P_{\text{PEO}}\phi_{\text{PEO}} + P_{\text{PS}}\phi_{\text{PS}} \tag{3.1}$$

where P_{Film}, P_{PEO}, and P_{PS} are the permeability values of composite film PEO and PS control, and ϕ_{PEO} and ϕ_{PS} are the fractions of PEO and PS in the composite structure, respectively [2,3].

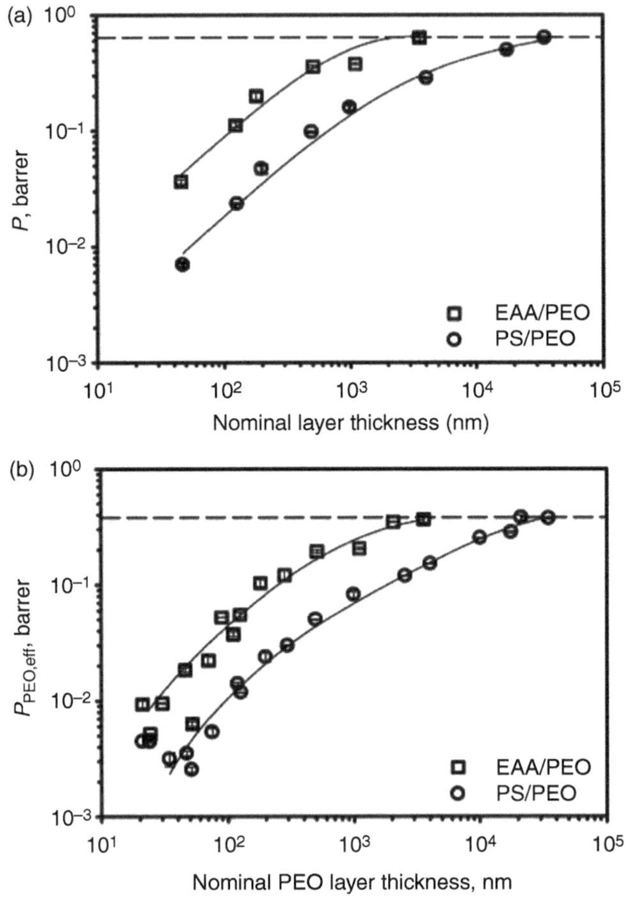

Figure 3.1 Effect of layer thickness on oxygen permeability at 23°C and 0% RH. (a) Oxygen permeability of PEO/PS (50/50) and PEO/EAA (50/50) films. The dashed line indicates permeability of film permeability calculated from the series model. (b) The effective PEO layer permeability $P_{PEO,eff}$ from films of varying composition. The solid lines are drawn as guides. The dashed line indicates PPEO.

The changes in the barrier properties were correlated with the crystallization of PEO layers under PS or EAA confinement. The multilayered films revealed layer thickness dependent PEO crystallization impacting the lamellar morphology, which affected the gas transport properties. Confinement of PEO showed transition from 3D spehrullitic morphology to flattened spherulites and to highly oriented lamellar structures as the layer thicknesses changed from micro- to nanoscale as shown in Figure 3.2. Truncated spehrullitic morphology was observed in layers as thick as 21 μm, with nucleation predominantly occurring within the PEO

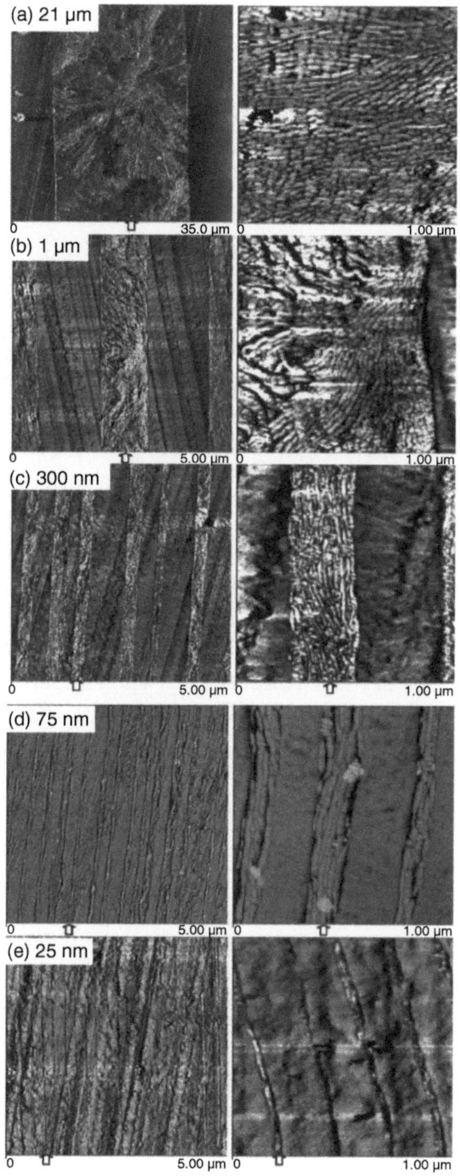

Figure 3.2 AFM phase images of partial cross-sections of PS/PEO-layered films. (a) 21 μm, (b) 1 μm, (c) 300 nm, (d) 75 nm, and (e) 25 nm PEO layers. Two magnifications of each film are shown. The images in the left column are 5 × 5 μm, except that the image in (a) is 35 × 35 μm; the images in the right column are all 1 × 1 μm. The arrows indicate a PEO layer. The lines in the PS layers were knife marks from the cyromicrotoming; otherwise the PS layers were featureless.

layers followed by the radial growth to spehrullitic morphology. Unlike bulk nucleation, flattened spherulites in 1 μm layers nucleated at the PS interface with constrained lamellar growth. The average diameter of the flattened- spherulite was close to 100 μm in diameter. Further reduction in layer thickness to 300 nm layers showed stacks of oriented in-plane lamellae. Some improvement in orientation of the stacked lamellae parallel to the layers was observed in 75 nm layers. Such morphology of polymers with chain orientation perpendicular to the layer surface, that is, lamellae parallel to the layer direction, is known as "in-plane" lamellar morphology. As the layer thickness approached 25 nm, close to PEO lamellar thickness, single-layered lamellae resembling the single crystal morphology of PEO were observed. Interestingly, even though the PEO was confined at few tens of nanometer thickness, the overall crystallinity of PEO decreased only slightly. The crystallinity of PEO control was measured as 76%, while 75 and 25 nm PEO layers showed crystallinity of 75 and 68%, respectively. As the crystallinity changed only slightly, the unusual drop in the permeability was, therefore, attributed to the changing PEO lamellar morphology. The PEO lamellar orientation was calculated using Herman's orientation function of the azimuthal intensity distributions of two-dimensional (2D) wide angle X-ray scattering (WAXS) of the PEO crystal reflections [4]. The Herman's orientation function values quantitatively classified different PEO orientations. Predominantly in-plane oriented lamellae with decreasing layer thickness were confirmed using this approach. In-plane lamellae in PEO nanolayers act as barrier to gas transport and, any defects such as lamellar edges provide diffusion pathway to gas molecules. Degree of lamellar orientation in PEO layers was strongly dependent on the confining substrate. The PEO orientation calculated from analysis of 2D WAXS patterns using Herman's orientation function revealed higher orientation of PEO lamellae in presence of PS layers than EAA layers for comparable layer thickness. Detailed effect of confining substrates on the lamellar orientation in nanolayers was investigated in polycaprolactone (PCL) based multilayered films, as discussed in the next part.

According to a two-phase model, crystalline phase act as a barrier to gas transport and amorphous part is a permeable component. In the layered composite, oriented lamellar crystalline morphology was impermeable to gases, while the lamellar fold surfaces consists of permeable amorphous region. When the lamellae are highly oriented, in-plane lamellae in layered assembly resemble dispersed platelets as described by Cussler et al. as shown in Equation (3.2) [5]. Cussler model correlates the effect of aspect ratio of impermeable flakes and concentration on the permeability.

$$P = P_m \left[1 + \frac{\alpha^2 \phi^2}{4(1-\phi)} \cos^2 \theta \right]^{-1} \tag{3.2}$$

where ϕ is the volume fraction of the platelets taken as PEO crystallinity, α is the aspect ratio and P_m is the permeability of the matrix, P is the permeability of the film and θ is the angle between the gas flux and the normal to the platelets. For highly oriented PEO platelets in the multilayered films, $\cos^2\theta$ is close to 1.The aspect ratio values calculated for thinner layers increased rapidly from 48 in 114 nm PEO layers to 265 in 25 nm. Using crystalline core thickness of 16 nm, the lateral dimension of the PEO lamellae was estimated as 4 μm in PEO/PS films, whereas PEO/EAA films estimated lateral dimensions as 2 μm. It is evident that with decreasing layer thickness and increasing aspect ratio, the lamellar edge defects reduced improving the barrier properties [3].

To understand the origin of the in-plane lamellae in PEO nanolayers, isothermal crystallization kinetics was investigated by selective melting of PEO layers below glass transition temperature (T_g) of PS [6]. A typical PEO melting endotherm of 65°C was lower than the PS T_g of 98°C as shown in Figure 3.3. A temperature window between $T_{m,PEO}$ and $T_{g,PS}$ allowed selective melting of only PEO layers within the rigid confinement of PS and controlled isothermal crystallization at different temperatures. In this melt recrystallization approach, PEO layers were melted by heating the films to 90°C, below PS T_g of 98°C, to ensure complete melting of PEO

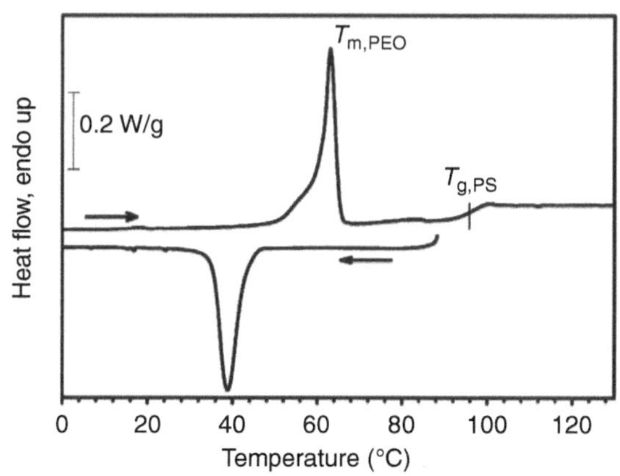

Figure 3.3 DSC thermogram of the PS/PEO 90/10 film with 1025 alternating layers and nominal PEO layer thickness of 25 nm. The heating curve was collected by heating a specimen from −10°C to 150°C. The cooling curve was collected by cooling another specimen after it was heated only to 90°C and held at 90°C for 5 min. The heating/cooling rate was 10°C/min.

layers. From crystallization kinetics data, the conventional chain folding in PEO layers with common side (σ) and fold surface energy (σ_e) product calculated as 3.1×10^{-4} J^2/m^4 in PEO nanolayers correlated well with the literature values of $(2.7–7.2) \times 10^{-4}$ J^2/m^4 [7]. However, significantly reduced crystallization rate with decreasing layer thickness suggested negligible impact of substrate nucleation on PEO layers. Heterogeneous nucleation within the PEO layers, therefore, was responsible for formation of in-plane lamellar crystals. Since the coextrusion processing conditions and materials were not changed during production trials, the nucleation density (number of nuclei per unit volume) of PEO remained constant in all films. However, as the PEO spread into micro- and nanolayers, the area nucleation density (number of nuclei per unit area) decreased by more than two orders of magnitude when the layer thickness changed from 110 μm to 25 nm. Therefore, the nucleation in the extruded films initiated at these heterogeneous nuclei produced truncated spherulites in thick layers and flattened spherulites with layer thickness reduction. Interestingly, flattened spherulites also showed significant nucleation at the PEO/PS interface. As the layer thickness approached few tens of nanometers, significant constrain on the PEO crystallization led to formation of oriented in-plane lamellae. The aspect ratio of the lamellae increased with decreasing layer thickness due to reduced area nucleation density. Such flattened spherulites and oriented lamellae, acted as barrier to gas transport in PEO nanolayers and the multilayered films [6].

It is obvious that the in-plane lamellae in PEO layers improve the gas barrier properties of the multilayered films. However, as noted in the extruded films, heterogeneous nucleation origin for PEO crystallization changed as a function of layer thickness. Thicker layers showed bulk nucleation of PEO layers, flattened spherulites revealed nucleation at PS interface and in-plane lamellae showed bulk nucleation. Further investigation of PEO crystallization temperature investigated using a controlled melt recrystallization approach as shown in Figure 3.3 showed an interesting phenomenon of transition of lamellar orientation with decreasing crystallization temperature in PEO layers. Below 20°C crystallization temperature, PEO preferably crystallized "on-edge" (lamellae perpendicular to the layer direction, chain axis parallel to the layers), followed by a sharp transition to in-plane crystallization above 25°C crystallization temperature. The transition was attributed to change in the nucleation mechanism from heterogeneous nucleation above 25°C to substrate-assisted nucleation below 20°C. At higher crystallization temperature, substrate effect is deactivated and PEO layers are nucleated heterogeneously within the bulk.

An example of on-edge crystal orientation in 75 nm PEO layers after melt crystallization at 20°C and comparison with as-extruded PEO layers

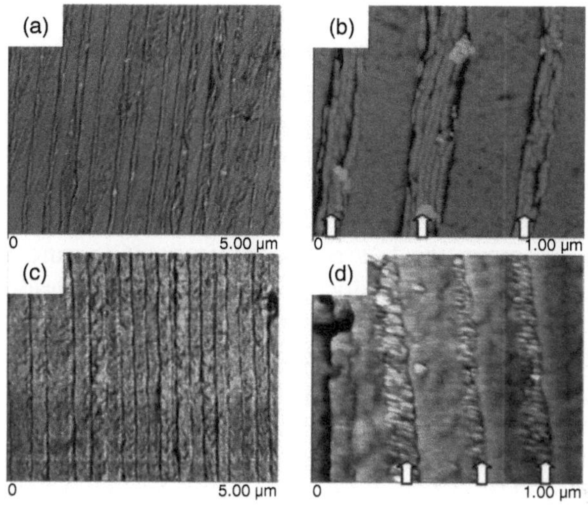

Figure 3.4 (a,b) AFM phase images of partial cross-sections of the extruded film with 75 nm-thick PEO layers. (c,d) AFM phase images of the film after melting and recrystallization by quenching of molten PEO layers to −20°C. The arrows indicate the PEO layers.

is shown in Figure 3.4 [8]. It is also noted that the extruded films were collected on water-heated take-off rolls during film extrusion process, which allowed the crystallization of PEO significantly above 25°C resulting into in-plane lamellar orientation. It is obvious that on-edge orientation should reduce the resistance to gas transport as vertical lamellae created significantly larger defects in PEO layers for increased gas permeability. Such transition from on-edge to in-plane lamellar orientation was investigated in more detail by confining PCL against different substrates. The effect of the lamellar orientation on the gas transport properties was investigated in sPP multilayered films [8].

To use PEO based multilayer films in commercial applications, it is necessary to understand the effect of humidity on the transport properties as PEO absorbs significant amount of water. PEO/PS films with 20 wt% PEO showed less than 2% moisture sorption and the equilibrium moisture absorption increased to 4% at 50 wt% PEO content. Importantly, oxygen permeability of PS/PEO layers changed only slightly in presence of 85% relative humidity.

3.2.2 Polycaprolactone (PCL)

Similar to PEO-based multilayered films, confinement of PCL in PCL/PS multilayered films at nanoscale showed a 200-fold improvement in the barrier properties when the number of layers increased from 17 to 1025 [9]. The corresponding PCL layer thicknesses in this set of films changed from 16 μm to 35 nm. PCL control films showed oxygen permeability of 0.97 ± 0.03 barrer, while PCL/PS (50/50, vol./vol.) films with 1025 layers resulted into 0.005 ± 0.001 barrer permeability. The effective PCL layer permeability showed 350-fold decrease in oxygen permeability in 35 nm layers as compared to 16 μm PCL layers in 17-layered films. Similar to PEO nanolayer confinement, the low permeability in PCL nanolayers resulted from the oriented lamellae. Reduced layer thicknesses gradually changed the morphology from isotropic spherulites in 16-μm thick layers to flattened/truncated spherulites in 2 μm layers to stacks of lamellae in 250 nm layers to very large in-plane lamellae, which resembled single crystal structure. The lateral dimensions of the oriented lamellae in PCL layers were estimated in the range of 0.75–1.6 μm, which act as impermeable platelets for gas diffusion. Using Cussler model (Equation (3.2)), estimated aspect ratio of PCL crystals was as high as 270 in 35 nm PCL layers [9].

Another set of PCL-multilayered films with PMMA as a confining layer showed a similar improvement in the barrier properties, which correlated well with the orientation function of PCL crystals as shown in Figure 3.5. Further investigation of PCL confinement in PCL/PC system resulted into complete miscibility of the components due strong interdiffusion during processing and absence of any confinement effects.

It is obvious that the interaction of confining substrates play an important role on the crystallization behavior of the confined material. Selecting a proper confining polymer is important in fabricating multilayered films for high gas barrier packaging applications. In addition to PS and PMMA, two other semi-crystalline polymers, nylon 12 (PA12) and polypropylene-graft-maleic anhydride (PP-*g*-MA), showed formation on in-plane lamellae in extruded films [10]. Isothermal melt crystallization showed on-edge to in-plane lamellar transition at 30°C in PCL/PA12 films and 45°C in PP-*g*-MA/PCL films. A significant shift in the PCL crystallization in PCL/PP-*g*-MA and PCL/PA12 films from on-edge to in-plane lamellae to higher temperatures suggested additional thermodynamic and kinetic phenomena during confined crystallization. The crystallization of PCL on semicrystalline polymer substrates has shown substrate induced transcrystalline morphologies of PCL in other research [11]. As the crystallization temperature is critical in determining the lamellar orientation in these systems, special attention to ensure the appropriate film collection temperature during co-extrusion processing is required. In case of PS and PMMA substrate, film collection on heated take-off rolls during the extrusion process allowed crystallization of PCL significantly above 10°C to form in-plane lamellae. On the other hand, PA12 and PP-*g*-MA confining layers produced on-edge PCL lamellae in the extruded films [10] (Figure 3.6).

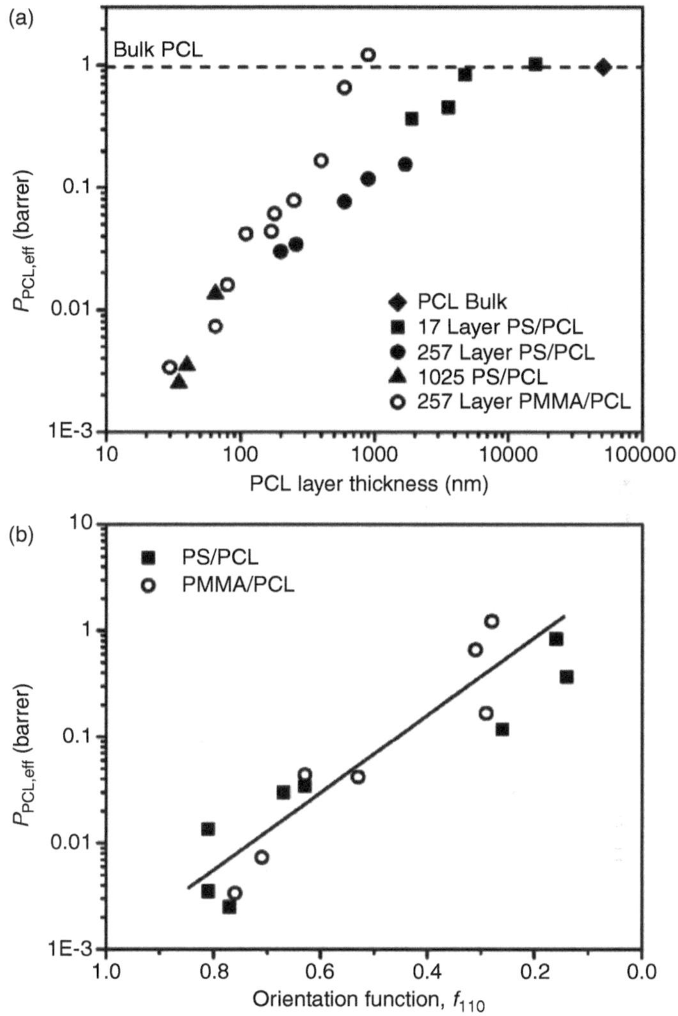

Figure 3.5 Effect of layer thickness on the effective permeability of the PCL layers. (a) The decrease in $P_{eff,PCL}$ as the layer thickness decreases from the microscale to the nanoscale and (b) the correlation between $P_{eff,PCL}$ and the orientation function for PS/PCL and PMMA/PCL. The solid line is drawn as a guide.

Ability of PEO and PCL to produce oriented lamellae in multilayered films in layer thicknesses below few hundred nanometers and the single-crystal like structures in 20–30 nm layers is unique in the coextruded films. The structural evolution in the PEO and PCL layers under confinement is shown schematically in Figure 3.7a. The oxygen permeability properties of the layered composites and the nanolayers showed its strong correlation with the lamellar orientation. Increased tortuosity to gas diffusion pathway

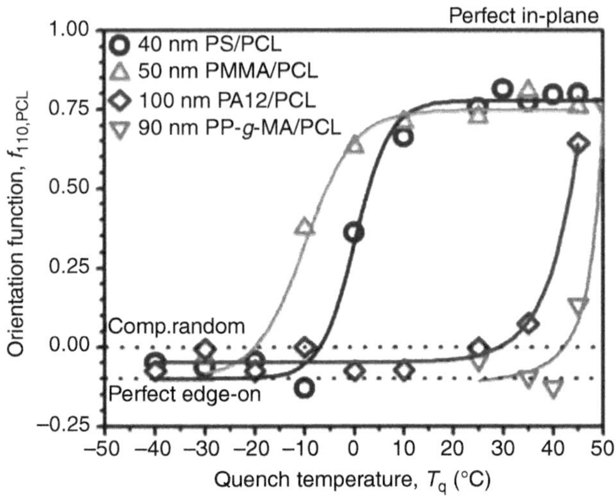

Figure 3.6 Herman's orientation function PCL f_{110} as a function of quench (isothermal melt recrystallization) temperature for all crystalline and amorphous systems, indicating the different transition regions in PCL crystal orientation from in-plane to on-edge as the quench temperature is decreased.

as a function of increased lamellar orientation and reduced layer thickness is schematically shown in Figure 3.7b [2,10].

3.2.3 Syndiotactic Polypropylene (sPP)

Confinement of sPP (T_m = 126°C) against amorphous glassy PC (T_g = 145°C), allowed sPP melt crystallization under the confinement of PC [12]. Morphological analysis of extruded film did not show any truncated

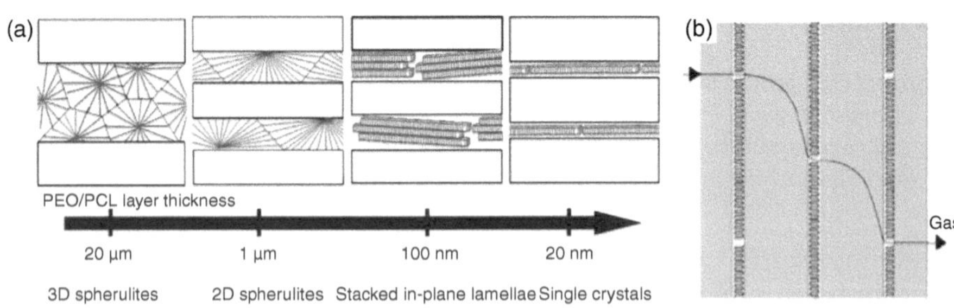

Figure 3.7 (a) Structure/morphology evolution of PEO and PCL confined layers, as the layer thickness is reduced from the micro- to nanoscale; (b) schematics of gas diffusion pathway through nanolayered PEO and PCL confined in multilayered films.

spherulites or in-plane lamellar structure in sPP nanolayers, as observed in PEO and PCL layers. The multilayered films showed randomly oriented sPP lamellae and reduced crystallinity, without significant effect on the gas transport properties. As discussed earlier, isothermal melt recrystallization of confined PEO and PCL nanolayers demonstrated lamellar orientation change as a function of temperature. This approach was further extended to selectively melt sPP nanolayers in 30/70 film compositions of 257-layered sPP/PC film, below the PC T_g of 145°C in sPP/PC multilayered films followed by isothermal melt recrystallization of 30 and 120 nm sPP layers in 60–105°C temperature range. Controlled melt recrystallization preferably formed in-plane sPP lamellae above 90°C and on-edge lamellae below 70°C crystallization temperatures, while intermediate temperatures showed mixed crystals populations. It is, therefore, obvious that the melt recrystallization of sPP layers during the film production process occurred at 70°C or lower temperatures, thereby preventing the formation of in-plane lamellae in extruded films. A subsequent step with carefully controlled crystallization temperatures allowed in-plane lamellae orientation. The effect of the crystallization temperature and time also explains the origin of randomly oriented crystals in as-extruded sPP/PC films. The crystallization half time of 5 min was measured for 30 nm sPP layers for isothermal crystallization at 89°C. In a typical film take-off process during extrusion process, faster quenching processes prevented formation of in-plane or on-edge lamellae due to shorter crystallization times. The sPP/PC films required postextrusion thermal treatment to achieve the lamellar orientation unlike PEO and PCL based films.

Oxygen transport measurements showed an order of magnitude decrease in the permeability of sPP layers with increasing melt crystallization temperature. Similar to PEO and PCL films, the reduced permeability correlated well with the in-plane lamellar orientation. As postulated earlier, the as the fraction of on-edge lamellae increased, the oxygen permeability increased significantly. Within the crystallization temperature range of 60–110°C, the oxygen permeability of sPP layers decreased by at least one order of magnitude, Figure 3.8. This change in permeability correlated well with the in-plane or on-edge crystal fractions as shown in Figure 3.9. In this case, the melt crystallization approach was successfully used to control the lamellar orientation in sPP layers under the rigid PC confinement. This approach can be extended for developing designed multilayer packaging films with desired gas transport properties [12].

A similar observation in regards to formation of randomly oriented crystals during film processing was made for isotactic polypropylene (iPP) in iPP/PC multilayered films [10]. Randomly oriented crystals and reduced crystallinity values were observed in iPP nanolayers, which are expected

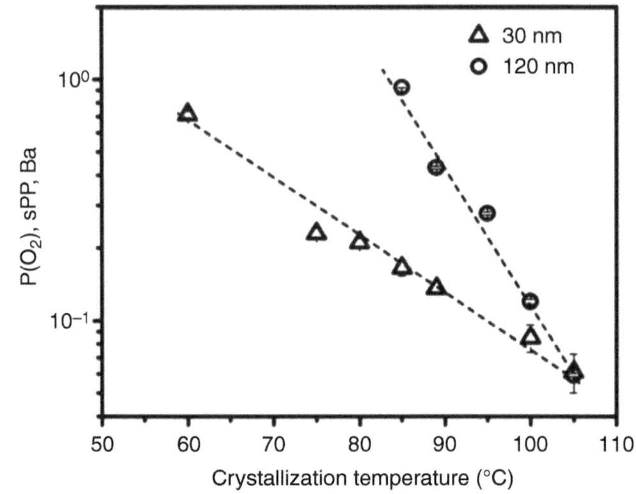

Figure 3.8 Effective permeability of the sPP layers as a function of crystallization temperature. The dotted lines are shown as guides.

to show no impact on the gas transport properties in iPP/PC multilayered films. An approach of isothermal melt crystallization showed some promise in producing oriented lamellae in iPP layers as controlled melt recrystallization of iPP layers at 140°C did not produce entirely in-plane lamellae for improved gas barrier properties. For example, iPP layers as

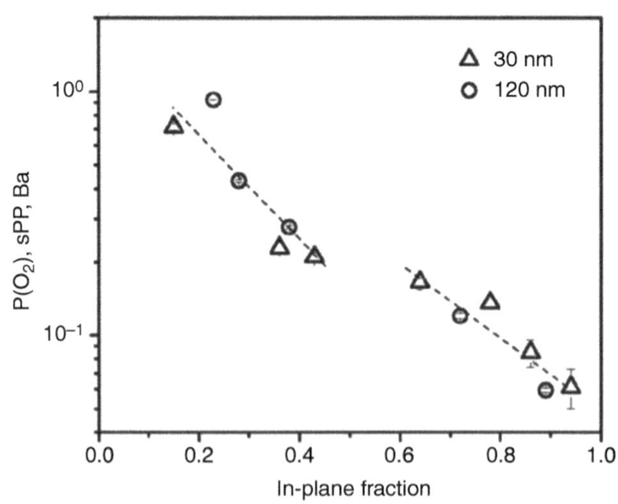

Figure 3.9 Correlation between P_{sPP} and the in-plane fraction calculated from WAXS patterns. The dotted lines are shown as a guide.

thin as 80 nm showed large fractions of in-plane lamellae and significant cross-hatching typical of iPP. Although in-plane lamellae would improve the barrier properties in iPP layers, the presence of cross-hatched lamellae can increase the gas permeability. Further investigation of iPP based multilayered films is required to evaluate the gas transport properties.

3.2.4 Polyvinylidene Fluoride (PVDF) Polymers

PVDF homopolymer and copolymers are known to have high barrier properties to gases and water. Multilayered films of PVDF and PVDF–TFE were coextruded with high T_g PC and PSU materials, which allowed melt recrystallization to control the lamellar orientation [13]. The extruded multilayered films of PVDF-based polymers did not show any preferred crystal orientation to impact the gas transport properties significantly. Therefore, controlled melt recrystallization of PVDF homopolymer was performed, at 135, 145, and 170°C, above the bulk crystallization temperature of 130°C. In PVDF–TFE layers, 115 and 120°C isothermal crystallization temperature was above the bulk crystallization temperatures of 110°C. The gas transport properties improvement of 100× was observed in 30 nm layers, as compared to 200–500 nm layers in the two polymers, while the overall permeability of PSU/PVDF and PC/PVDF–TFE films improved by a factor of 25–30 as shown in Figure 3.10. Using Cussler model, increased aspect ratio of oriented lamellae was estimated with increasing crystallization temperature and decreasing layer thicknesses. Similar to earlier discussed films, in-plane crystals preferentially formed at higher isothermal melt recrystallization temperatures, while on-edge crystals formed at low crystallization temperature. Oxygen permeability

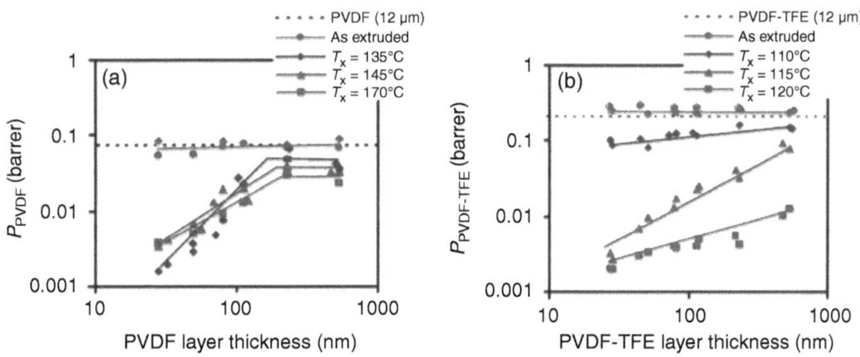

Figure 3.10 Effective oxygen permeability of PVDF-based layers versus layer thickness. (a) PVDF layer permeability in the PSU/PVDF films and (b) PVDF–TFE layer permeability in the PC/PVDF–TFE films.

values as low as 0.005 barrer, were measured for 30 nm PVDF nanolayers. Increased gas barrier in PVDF multilayers also implies possible high barrier in these films to other chemicals and solvents [13].

3.2.5 High Density Polyethylene (HDPE)

Multilayered films of HDPE/PC and HDPE/PS showed twisted HDPE lamellae in layers as thin as 30 nm [14]. Unlike the multilayer film systems discussed earlier, isothermal melt recrystallization in HDPE/PC system approach did not change the morphological features in HDPE nanolayers significantly; preventing large improvement in the barrier properties of HDPE based multilayered films. However, melt recrystallized HDPE layers showed a twofold improvement in the oxygen barrier properties and twofold improvements in water barrier properties due to twisted lamellae and increased crystallinity [15]. Oxygen barrier in extruded PC/HDPE (90/10, 40 μm) film was 2.9 ± 0.2 barrer, while recrystallization at 128°C showed oxygen permeability of 1.2 ± 0.1 barrer. The water permeability also decreased from 5.3 ± 0.2 g mil/100 in.^{2} day to 1.8 ± 0.2 g mil/100 in.^{2} day.

3.2.6 Poly(4-methylpentene-1) (P4MP1)

Multilayer film confinement approach to melt recrystallize P4MP1 resulted into high aspect ratio, in-plane lamellar crystals under the confinement of syndiotactic polystyrene (sPS) and polyphenyline sulfide (PPS) blend. P4MP1 is a unique material with crystal phase density lower than its amorphous phase density [16]. This low density was attributed to the crystal polymorphism in P4MP1, where 4.5 Å channels between the methylene and methane groups of the P4PM1 molecules run along the *c*-axis. Therefore, the crystalline regions in this polymer can permeate gas molecules smaller than 4.5 Å. If P4MP1 chains are perpendicular to the gas flux, the in-plane crystals formed can allow increased gas flux through P4MP1 layer and increase the gas permeability of the layers. Using an approach of melt-recrystallization of P4MP1 layers while creating the porous confining substrate, improved oxygen permeability was demonstrated in P4PMP1 based multilayered films.

The melting temperatures of the sPS and PPS of 274 and 278°C allowed for selective melting of P4MP1 below 270°C and controlled crystallization at 215°C. To increase the confining layer permeability, the blend layer was stretched prior to P4MP1 melt recrystallization thereby creating multiple pores in the sPS-PPS blend layer and increasing its permeability. Porous confining layers allowed accurate measurement of the gas transport properties of P4MP1 layers. The measured permeability values of P4MP1 layers showed a 3× improvement in the oxygen transport due to alignment of P4MP1 channels along the *c*-axis perpendicular to the film surface. The channel diameter of 4.5 Å is significantly larger than the oxygen, CO_2, H_2 diameter; however, less than methane molecules. Alignment of the open channels reduced hindrance to gas diffusion and increased the gas permeability. Interestingly, when the P4MP1 crystals are randomly oriented, the barrier to gas transportation is higher than aligned lamellae. This polymer can be used as a gas separation membrane.

3.2.7 Novel Composites From Multilayered Structures

Development of PEO, PCL, sPP, and PVDF-based multilayered films showed dramatic improvements in the gas barrier properties of commercial polymers. In-plane crystallization of these polymers imparted more than two orders of magnitude reduction in oxygen permeability. A unique opportunity to use PEO and PCL nanolayers in commercial packaging materials such as biaxially oriented PP was developed. Another system with PVDF–TFE as a high barrier material in PET was also investigated for biaxial stretching capabilities to improve the final film barrier properties.

3.2.7.1 Nanobiaxially Oriented Polypropylene (Nano-BOPP)

Biaxially oriented PP (BOPP) film is widely used in flexible packaging applications due to its improved toughness, transparency, and oxygen barrier properties over unoriented PP [17]. Various fillers with high aspect ratio have also been investigated to improve the oxygen barrier in BOPP films at the expense of optical properties [18]. Using PEO or PCL nanolayers in BOPP film at thicknesses lower than the quarter wavelength of light allows barrier improvement without sacrificing the mechanical and optical properties. Multilayered sheets with 33-layered PP/PEO structure were produced by layer multiplication process followed by addition of thick PP skin layers in the coextrusion process [19]. The final composite structure was produced in $PP[(PEO/PP)_n]PP$ form with significantly thick PEO layers (2.5 μm). To improve the barrier properties, it was necessary to reduce PEO layer thickness down to a few tens of nanometer. Biaxial stretching of the composites produced oriented PP film and achieved simultaneous reduction in the PEO layer thickness to nanoscale by its melting and recrystallization into in-plane lamellae. The schematic of this process transforming micro- to nanolayers is shown in Figure 3.11a. Multiple processes – coextrusion followed by biaxial stretching and melt recrystallization–created novel BOPP films with PEO nanolayers (nano-BOPP).

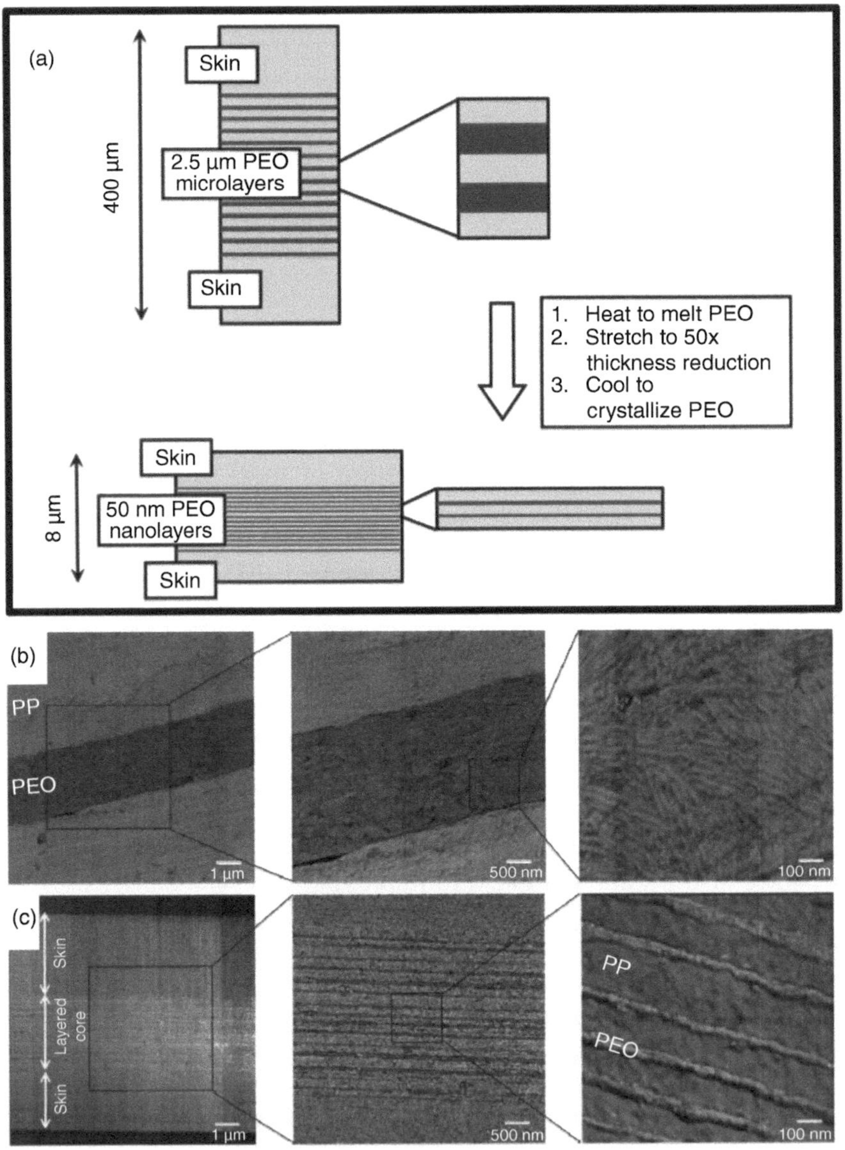

Figure 3.11 (a) Schematic showing the biaxial stretching of microlayers to nanolayers to produce nano-BOPP film, cross-section AFM images of PP/PEO-90/10 showing the layer morphology; (b) before stretching (1 × 1); and (c) after biaxial stretching (7 × 7) at 145°C.

Biaxial stretching of the composites was carried out in 135–150°C temperature range to draw ratios of 5 × 5, 6 × 6, and 7 × 7. An example of cross-section image of the composite sheet with PP/PEO (90/10, vol./vol.) as a core structure is shown in Figure 3.11b. A thick PEO layer visible in the image shows randomly oriented PEO lamellae. Biaxial stretching to 7 × 7 draw ratio produced nano-BOPP films with 16 PEO layers in 33-layered PEO/PP core. Individual PEO layers down to 38 nm were successfully produced with in-plane lamellar structures. The oxygen permeability 0.703 ± 0.041 barrer in BOPP control films reduced to 0.029 ± 0.007 barrer in nano-BOPP film with 10% PEO after stretching to 7 × 7 draw ratio. The effective PEO permeability decrease was similar to earlier work. The multilayered films also showed a CO_2/O_2 selectivity of 10, when PEO layers were present in the core. Transparency properties did not show any significant changes in presence of PEO as the PEO layer thicknesses were significantly below the quarter wavelength of light. An improvement of 10% in tear energy was measured in nano-BOPP films. Because the properties of PEO are significantly affected in presence of water, measurement of permeability properties in presence of high humidity was performed. The oxygen permeability of nano-BOPP films showed a factor of two increment in effective permeability in presence of 85% relative humidity. However, the overall film performance was still significantly better than the neat BOPP films.

This approach was further extended to using PCL as a barrier layer in the multilayered BOPP composite. Multilayered films with a layered core structure of PCL/PP and outside thick layers of PP were produced in PP[(PCL/PP)$_n$]PP format. The PCL volume fraction range varied from 3% to 10% [20]. The multilayered films were biaxially oriented at 155°C, which was sufficient to melt PCL nanolayers (T_m = 57°C). Subsequent cooling of multilayered films to room temperature, however, resulted into formation of on-edge lamellae in presence of PP due to transcrystallization at the interface. The film structure was modified to incorporate PS layers as a substrate layer in the film structure to produce PP[(PP/PS/PCL/PS/PP)$_n$]PP film, where PCL layers were confined against PS layers. Biaxially stretched BOPP and melt recrystallized PCL nanolayers improved the gas transport properties by at least one order of magnitude, from 0.635 ± 0.018 barrer in BOPP to 0.063 ± 0.002 barrer in nano-BOPP with 15% PCL. Processing conditions were optimized to isothermally quench the films between 33 °C and 41°C for improved barrier, as the air-quenched samples did not show significant improvement. Further analysis of the films for mechanical properties (tear energy and puncture strength) did not show significant changes with the addition of PCL and PS components in the layered structure.

3.2.7.2 Nanobiaxially Oriented PET (Nano-BOPET)

Biaxially oriented PET films are widely used as high barrier packaging material with improved mechanical and barrier properties [21]. Combining the confinement effect of PVDF–TFE layers, the improvement in the barrier properties was demonstrated in biaxially stretched PET/P(VDF–TFE) films [22]. Thirty-two- and 256-layered PET/P(VDF–TFE) films with wide range of compositions, and layer thicknesses ranging from 10 μm to 24 nm were fabricated. Measured oxygen permeability of the extruded PET/P(VDF–TFE) films was close to the series model prediction suggesting that the inherent morphology of PET and PVDF–TFE layers was similar to monolayer films. Multilayered films with thick layers were biaxially stretched to reduce the individual layer thicknesses and achieve orientation for improved mechanical and optical properties. Biaxial stretching of the films exhibited on-edge lamellae formation of confined PVEF–TFE layers and the oriented PET layers showed stacked fibrillar structures resulting into overall oxygen permeability reduction by a factor of 2–3. Stacked fibrillar crystals of PET formed during biaxial stretching imparted the barrier improvement in the nano-BOPET films. Using an approach of melt recrystallization of PVDF–TFE nanolayers at 120°C, high aspect ratio in-plane lamellae were obtained in these films and the overall barrier properties of the films improved by an order of magnitude in the biaxially stretched and melt recrystallized films as shown in Figure 3.12. The effective permeability of PVDF–TFE layer estimated from a series model showed a 400-fold reduction over as-extruded PET/P(VDF–TFE) films. Similar to oxygen barrier improvement, the nano-BOPET films also showed a factor of 2–3 improvement in water vapor transport rates (WVTR) with an effective PVDF–TFE nanolayer improvement of 100- to 150-fold in biaxially oriented and melt recrystallized layers, as shown in Table 3.1 [22].

3.2.7.3 Filled Multilayered Film Systems

Incorporation of high aspect ratio inorganic particles in composites is known to improve the gas barrier properties in conventional polymers [23]. The shape and size of the inclusions, the volume fraction, interaction with polymer matrix and the alignment of the particles play an important role in improving the barrier properties. Alignment of the inorganic platelets perpendicular to the gas diffusion increases the tortuosity to gas transport and hence improve the barrier significantly. However, inclusion of high volume fraction of inorganic particles results into brittle and translucent films.

Multilayer coextrusion process was used to fabricate films with inorganic phosphate glass (P_{glass}) spherical particles incorporated in alternating

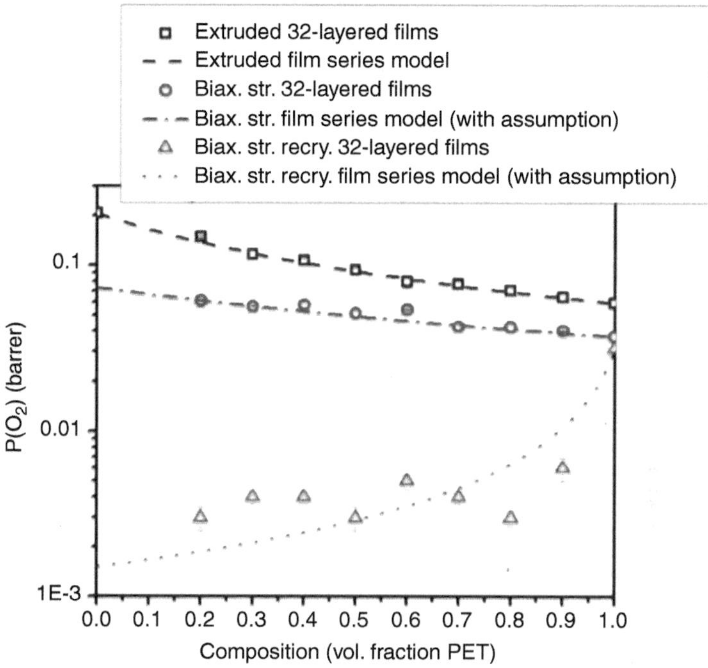

Figure 3.12 $P(O_2)$ of extruded, biaxially stretched, and biaxially stretched recrystallized PET/P(VDF–TFE) 32-layered films at various compositions plotted against the series permeability model describing permeability of layered systems.

layers of PP-*g*-MA/(PP-*g*-MA + P_{glass}) multilayered films [24]. When the films were biaxially stretched at 155°C, significantly above the P_{glass} T_g of 120°C, P_{glass} fillers translated into high aspect ratio oriented platelets as shown in Figure 3.13. The film cross-section resembled alternating layers of PP-*g*-MA and P_{glass} resulting into the oxygen permeability decrease of two to three orders of magnitude. The reduced oxygen permeability in the multilayered films was explained using Nelson and Cussler models. As the volume fraction of P_{glass} increased in the layers, the overlap of glass platelets increased the tortuosity of the diffusion pathway changing the mechanism from Nielsen to Cussler. When the P_{glass} platelets have the aspect ratio α oriented perpendicular to the gas flux and the total P_{glass} fraction in the composite is ϕ, the Nielsen and Cussler permeability model is given by Equations (3.3) and (3.4), respectively.

$$\frac{P}{P_m} = \frac{1-\phi_{P_{glass}}}{1-\alpha(\phi_{P_{glass}}/2)} \tag{3.3}$$

Table 3.1 **WVTR of Extruded, Biaxially Stretched, and Biaxially Stretched Recrystallized PET/P(VDF–TFE) Films**

Number of Layers	Composition PET/ P(VDF–TFE)	PVDF–TFE Layer Thickness (nm)	Overall Film WVTR (gm mil/100 in.^{2}day)		
			Extruded	Biaxially Stretched	Recrystallized Biaxially Stretched
1	PET control	–	1.35 ± 0.03	0.81 ± 0.07	0.66 ± 0.09
	PVDF control	10,000	1.09 ± 0.19	–	–
32	20/80	500	0.95 ± 0.06	0.37 ± 0.03	0.03 ± 0.01
	80/20	120	1.34 ± 0.06	0.68 ± 0.02	0.04 ± 0.01
256	50/50	40	1.05 ± 0.02	0.58 ± 0.02	0.02 ± 0.01
	70/30	24	1.35 ± 0.03	0.73 ± 0.02	0.04 ± 0.01

Selective data from Table 3 from Ref. [22].

$$\frac{P}{P_{\mathrm{m}}} = \frac{1-\phi_{P_{\mathrm{glass}}}}{1-\left(\alpha(\phi_{P_{\mathrm{glass}}}/2)\right)^2} \tag{3.4}$$

where P, and P_{m} are the permeability values of the film and the PP-*g*-Ma matrix, respectively.

Cussler model accounts for the overlapped P_{glass} platelets and reduced cross-sectional area between the platelets. The data fit using Equations (3.3) and (3.4) in Figure 3.14 showed that the Cussler model explains the

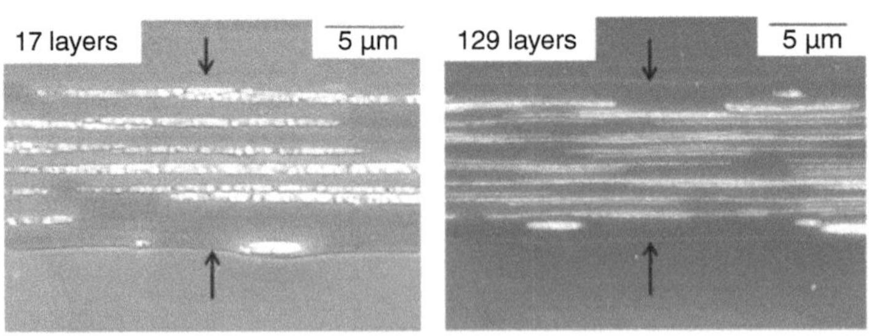

Figure 3.13 SEM images of biaxially oriented microlayered film with 10 vol% P_{glass} across the film cross-section. The arrows define the cross-section of the film.

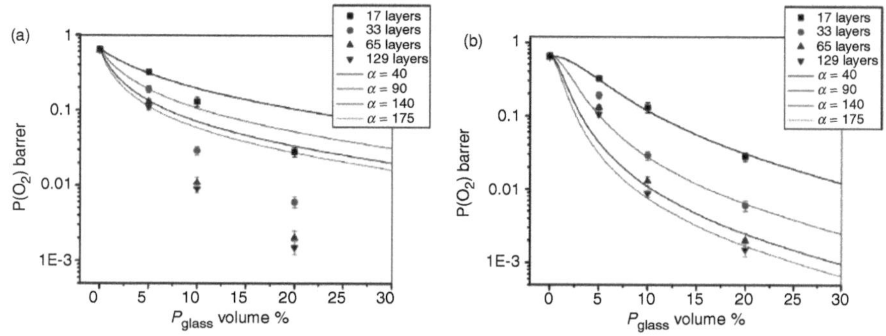

Figure 3.14 Experimental data for oxygen permeability versus volume fraction of P_{glass} for biaxially oriented composite films with various fits. According to (a) Nielsen and (b) Cussler equations.

data well. In overlapped P_{glass} platelets, repeated scattering of the gas molecule in neighboring platelets led to increased tortuosity and eventually increased barrier. An improvement of over two orders of magnitude, from 0.62 ± 0.02 barrer in PP-*g*-MA control to 0.002 ± 0.0005 barrer in PP-*g*-MA with 20 vol% of P_{glass}, was observed in 65-layered biaxially stretched films. The length of the P_{glass} platelets in the layers was approximately 15–20 μm in all the films, while the thickness decreased from 350 nm in 17 layers to 80 nm in 129 layers. Improvement in the gas barrier was accompanied by a twofold modulus increase, good flexibility, optical transparency up to 85%, and the films also maintained excellent performance under humid conditions [24].

3.2.7.4 Injection Molding Of Multilayered Films

An approach to create micro-platelets from the layered films, by selective melting of one of the layers was used to improve the barrier properties. The multilayered composites of PP and polyamide-66(PA66) were extruded with average layer thicknesses down to 500 nm [25]. Multilayers of PP/PA66/PP-*g*-MA (65/25/10) were produced with 33, 257, and 2049 number of layers. A melt blend of the three components was also prepared for comparison. The injection molding of the PP/PA66/PP-*g*-MA diced multilayered composites was carried out at 180°C, above PP melting and below PA66 melting temperatures. Selective melting of PP layers, while retaining PA66 layers resulted into high aspect ratio PA66 platelets in injection molded parts, which showed a four- to five-fold improvement in the barrier properties over the conventional melt blends of PP and PA66. Morphology analysis of the injection molded parts showed highly oriented PA66 platelets parallel to the surface of

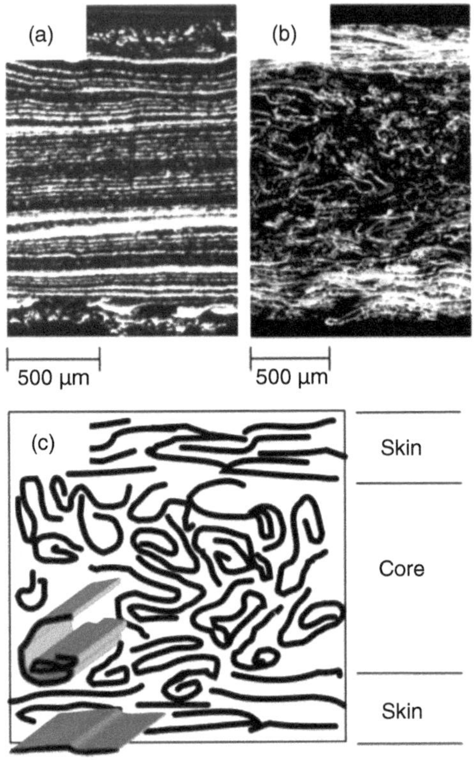

Figure 3.15 Blend morphology. (a) Microlayer with 257 layers; (b) 257-layered injection molded; and (c) schematic representation of injection molded microlayer.

the molded parts, while the core sections showed coiled platelets with preferential orientation perpendicular to the injection directions due to complex flow during injection molding negatively impacting the gas transport properties [26]. The morphology of two sections of the injection-molded parts is shown in Figure 3.15. If the orientation of the PA66 platelets is maintained parallel to the surface of the in the injection molded parts, further improvement in the gas barrier properties is possible. This approach can be used in multilayered film systems with significant differences in the melting temperatures of the component polymers [25] (Table 3.2).

Development of nanoconfined multilayered systems with hundreds of layers showed true impact of the nanostructures in the commercial polymers. Significant improvements in the gas barrier properties in the multilayer film and composite systems discussed in this section are summarized in Table 3.3.

Table 3.2 Permeability Properties of Controls and Blends (cm^3(STP) cm/ m^2/day/atm)

Material	O_2 Permeability	CO_2 Permeability
PP injection molded at 180°C	7.5 ± 0.1	26 ± 2
PA66 as-extruded	0.068 ± 0.005	0.28 ± 0.02
33-layered as-extruded	0.30 ± 0.03	1.11 ± 0.02
33 layers injection molded at 180°C	1.08 ± 0.08	3.8 ± 0.2
257 layers injection molded at 180°C	0.96 ± 0.03	3.0 ± 0.1
2049 layers injection molded at 180°C	0.99 ± 0.02	3.6 ± 0.1
257 layers injection molded at 280°C	4.0 ± 0.1	
Melt blend compression molded at 280°C	4.4 ± 0.1	15 ± 1

Selective data from Tables 2 and 3 from Ref. [25].

3.3 Adhesion and Mechanical Properties of Multilayered Composites

Development of multicomponent polymer composites is driven by the ability to combine two or more polymers to achieve synergistic effect in the final product. Considerable interest in improving the mechanical properties of the polymers has led to the development of different approaches to combine polymers. The phase behavior of the individual components determines the physical properties of the composites. For example, use of elastomeric components in the brittle amorphous or semicrystalline polymers to improve the toughness is known. Rubber particle size, adhesion, and surface modifications are some of key parameters significantly affecting the deformation processes during failure [27]. Other approaches include addition of inorganic fillers, nanoparticles, additives, etc. Many heterogeneous systems include block or random copolymers, which can undergo intramolecular phase separation to produce ordered phases resembling multicomponent composites. For heterogeneous composites, many parameters affecting the final composite structure include nature of the components, the size and shape of the dispersed phase, interfacial interaction, blending conditions among many others. Multilayered films and composites provide an approach to create complex ordered micro- and nanoscale structures by combining two or more polymers. As the layer thicknesses approach few nanometers, the interface contribution to the mechanical properties becomes significantly large. In this section, a wide range of multilayered composites and the effect of layer thicknesses on the delamination toughness, failure mechanism, impact properties, ballistic properties,

Table 3.3 Summary of Various Multilayered Film Systems and their Improvements Over Control Films

Systems (Confined Layer/Confining Layer)	Compositions (vol./vol.)	Number of Layers	Film Thickness (μm)	Confined Layer Thickness (nm)	Film Permeability (Barrer)	Confined Layer Permeability (Barrer)	References
PEO control			127		0.38 ± 0.01	0.38	[3]
PS control			127		2.08 ± 0.01		
	50/50	9	142	18,000	0.50 ± 0.01	0.2842	
	50/50	33	130	4,000	0.283 ± 0.003	0.1518	
	50/50	257	51	200	0.047 ± 0.001	0.0238	
PEO/PS	50/50	1,025	48	114	0.0070 ± 0.0003	0.0035	
	80/20	1,025	21	21	0.0221 ± 0.0022	0.0045	
	90/10	1,025	125	25	0.0436 ± 0.0005	0.0089	
PCL control			51		0.970 ± 0.030	0.970	[9]
PS control			51		2.200 ± 0.200		
	50/50	17	250	16,000	1.397 ± 0.001	1.023	
PCL/PS	50/50	257	51	200	0.059 ± 0.006	0.03	
	50/50	1,025	37	35	0.005 ± 0.001	0.003	
PCL control			51		0.970 ± 0.030	0.970	[9]
PMMA control			51		0.057 ± 0.001		
	75/25	257	154	900	0.200 ± 0.001	1.221	
PCL/PMMA	25/75	257	86	170	0.053 ± 0.005	0.044	
	10/90	257	38	30	0.022 ± 0.003	0.003	
sPP control			50	50,000		1.181	[12][c]
sPP/PC as-extruded	30/70	257	50	120		1.460	
sPP/PC (T_c = 85°C)	30/70	257	50	120		0.940	
sPP/PC (T_c = 105°C)	30/70	257	50	120		0.060	

PVDF control			12	12,000	0.072	0.072	[13][c]
PVDF/PSU (as-extruded)	30/70		12	28		0.072	
PVDF/PSU (T_x = 135°C)	70/30	32	12	225		0.046	
PVDF/PSU (T_x = 135°C)	30/70	256	12	28		0.0017	
PVDF–TFE control			12	12,000	0.216	0.216	[13][c]
PVDF–TFE/PC (as-extruded)	30/70		12	28		0.269	
PVDF–TFE/PC (T_x =120°C)	70/30	32	12	225		0.006	
PVDF–TFE/PC (T_x = 120°C)	30/70	256	12	28		0.002	
HDPE control					0.23 ± 0.01		[15][c]
PC control					1.3 ± 0.1		
HDPE/PC (as-extruded)	10/90	257	40			2.9 ± 0.2	
HDPE (T_x = 128°C)	10/90	257	40			1.2 ± 0.1	
Nano-BOPP films							[19]
BOPP ($\lambda = 7 \times 7$, T_d = 145°C)			8		0.528 ± 0.020		
BOPP ($\lambda = 7 \times 7$, T_d = 145°C, T_x[b])			8		0.488 ± 0.012		
nano-BOPP/PEO ($\lambda = 7 \times 7$, T_d = 145°C)	90/10	33[a]	8	38	0.068 ± 0.007	0.0077	
Nano-BOPP/PEO ($\lambda = 7 \times 7$, T_d = 145°C, T_x[b])	90/10	33[a]	8	38	0.029 ± 0.004	0.0037	
Nano-BOPET films							[22]
PET ($\lambda = 4.5 \times 4.5$)			10		0.037 ± 0.001		
PET ($\lambda = 4.5 \times 4.5$, T_x = 120°C)			10		0.031 ± 0.002		
PVDF–TFE/PET ($\lambda = 4.5 \times 4.5$)	30/70	256	10	24	0.044	0.036	
PVDF–TFE/PET ($\lambda = 4.5 \times 4.5$, T_x = 120°C)	30/70	256	10	24	0.001 ± 0.0003	0.0004	

T_x is the confined layer recrystallization temperature; T_d is the biaxial stretching temperature; λ is the biaxial draw ratio.
[a]*The films constituted 33 core structure of PEO and PP with two thick skin layers of BOPP.*
[b]*Slow cooling of PEO layers performed as a recrystallization step. BOPP control sample was also cooled similar to PEO containing nano-BOPP films.*
[c]*The confined layer oxygen permeability values are estimated from the figures discussed in the corresponding references. These values are not reported in a tabular form in the references.*

tensile properties, deformation, and fatigue are discussed. Insertion of a tie layer to improve the compatibility and the delamination toughness of two polymers has also been demonstrated using coextrusion process. Recent work discusses use of gradient structure for controlling the failure behavior and effect of BCP confinement on the mechanical properties. Additional developments include controlling the lamellar orientation in submicrometer-thick layer by confined crystallization of polymers. These systems have provided new insights on the mechanical behavior of nanostructure materials. Identifying the property enhancements in composites with hundreds of layers provide an opportunity to develop practical applications beyond the currently used multilayered systems, which use only a few layers.

3.3.1 Interfacial Adhesion in Multilayered Films

Interfacial adhesion of the constituent polymers in the coextruded structure determines the delamination properties and hence the toughness of the composites. Delamination studies of multilayered composite system of PC and styrene-acrylonitrile copolymer with 25% acrylonitrile (SAN25) films showed layer thickness dependent modes of failure [28]. The number layers in the composite structure varied from 49 to 1857 and the PC/SAN compositions ranged from 26/74 (wt/wt) to 90/10 (wt/wt). The average layer thicknesses in these films were in 37–0.5 μm range. Delamination characterization was performed using T-peel test, where the specimens were notched using a fresh razor blade and ensuring the crack formation in single layer. It was revealed that the delamination mechanisms and the peel strengths were strongly dependent on the relative thicknesses of PC and SAN layers in multilayered films. Effect of relative layer thicknesses of the two components on the delamination toughness, which demonstrated four modes of failure, is shown in Figure 3.16. Single layer interfacial delamination showed debonding at the crack tip at either interfaces of SAN layer, followed by wrinkled layer formation and breaking of SAN25 layer. Multiple interface delaminations were observed when both SAN and PC layers were close to or below 1 μm thickness. Debonding ahead of crack tip, initial tearing through SAN layer, and subsequent tearing through multiple PC and SAN layers was observed in this mode of failure. In composites with thick SAN and thick PC layers, single layer crazing with multiple horizontal and vertical crazes ahead of the crack tip, and crack propagation through SAN25 crazes was observed. As the PC thickness reduced, the stress at the neighboring initiated the secondary cracks causing crazing in multiple SAN layers led to multiple layer craze delamination mechanism [28].

Multilayered composites with thin SAN layers were further investigated to measure the effect of acrylonitrile (AN) content on the interfacial

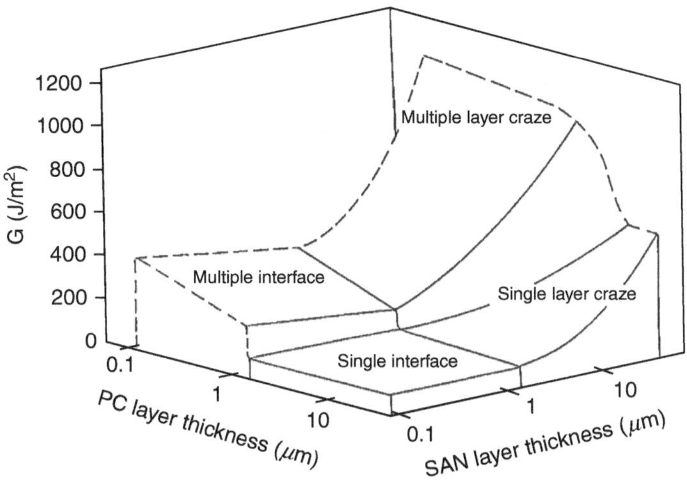

Figure 3.16 A 3D plot of peel toughness as a function of PC and SAN layer thicknesses.

toughness of the PC/SAN composites [29]. Multilayered composites of PC/SAN (1- to 2-mm thick) with wide range of layer thicknesses and varying amount of AN content were investigated. Delamination toughness of multilayers varied as a function of AN content and the layer thickness, as shown in Figure 3.17a. Based on failure analysis, three regions with different delamination mechanism were identified in these composites. At low AN concentrations and above AN concentration higher than 25%, the interfacial delamination was a preferred failure mode while crazing phenomena was predominantly observed in SAN layers with 15–25% AN. SAN materials with 20% AN content showed the highest delamination toughness in PC/SAN composites. The delamination toughness properties were associated with the two types of failure mechanism in the composites- interfacial failure and crazing failure. In general, SAN controls showed resistance to crazing with increasing AN content. A linear increase in critical mean stress, or craze condition, observed for SAN controls was defined as a stress required for internal craze growth in SAN controls. It was noted that the crazing condition is a critical parameter in defining the failure mechanism. Crazing failure occurred when the interfacial toughness of PC/SAN composites exceeded the SAN crazing conditions as shown in Figure 3.17b. Delamination toughness was higher than crazing condition when AN content was 15, 20, and 25% [29].

Improved delamination toughness was also observed in PC/PMMA multilayer composite, which showed an order of magnitude toughness increase over comparable PC/SAN25 composite, as shown in Figure 3.18 [30]. Examples of PC/PMMA (4.0/0.4 μm) and PC/SAN (2.5/0.5 μm)

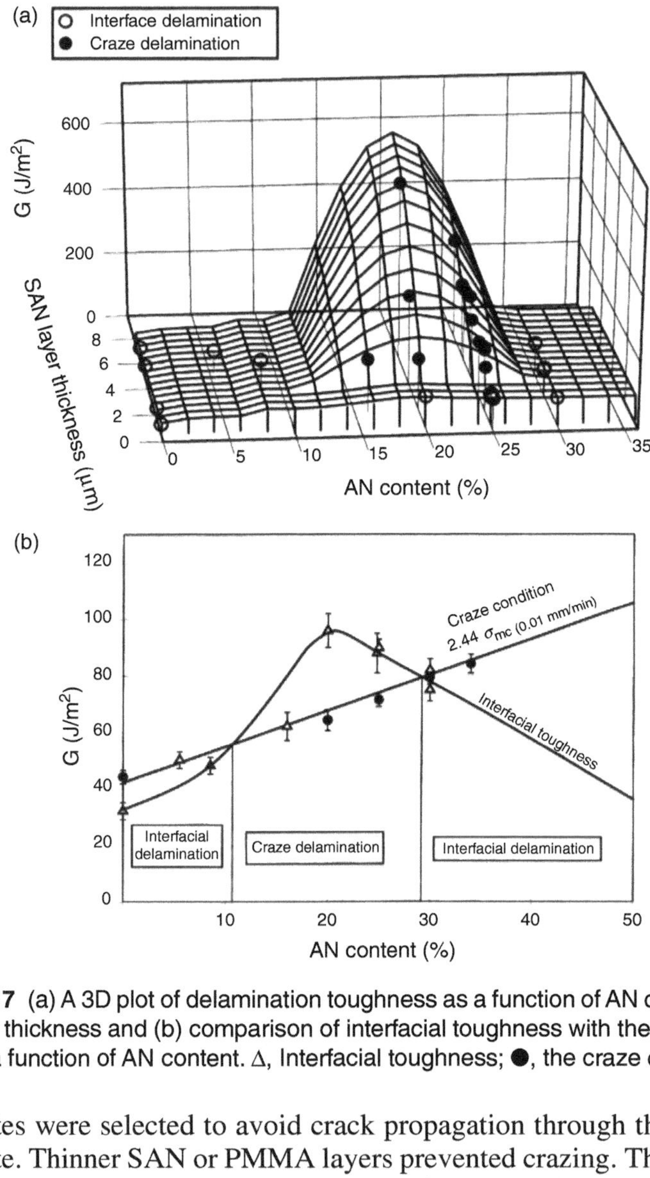

Figure 3.17 (a) A 3D plot of delamination toughness as a function of AN content and SAN layer thickness and (b) comparison of interfacial toughness with the craze condition as a function of AN content. Δ, Interfacial toughness; ●, the craze condition.

composites were selected to avoid crack propagation through the layered composite. Thinner SAN or PMMA layers prevented crazing. The delamination toughness of PC/PMMA multilayered tape was 950 J/m^2 and PC/SAN25 multilayered system was in the range of 70–90 J/m^2 for comparable layer thicknesses. Although PC and PMMA materials are not very compatible, it was argued that the PC/PMMA interaction is only weakly unfavorable for mixing [31]. Even with sharp interfaces in PC/PMMA composites, excellent adhesion improved the delamination toughness as well as impacted the stress–strain behavior of the composites [30].

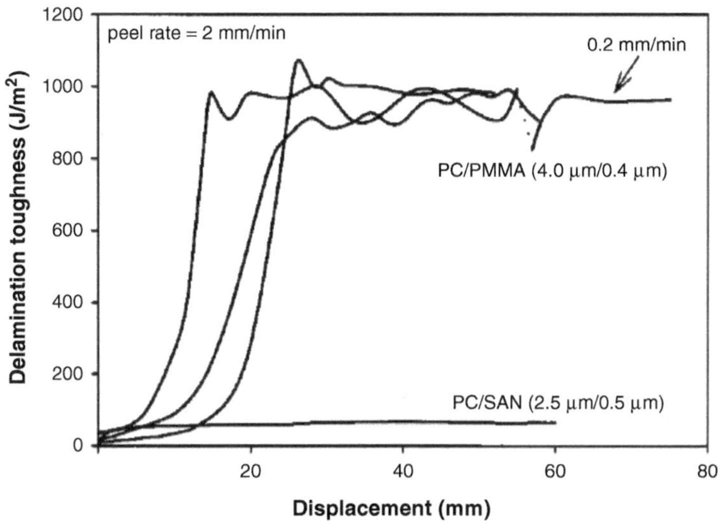

Figure 3.18 Peel curves of PC/SAN (2.5/0.5 μm) and PC/PMMA (4.0/0.4 μm) composites.

This approach was further extended to study adhesion properties of commercial copolymers. A series of ethyelene–styrene (ES) copolymers with variable styrene content, useful in compatibilizing PE and PS, developed by Dow Chemical Company's INSITE™ technology were coextruded with LDPE [32]. Different grades of ES shown in Table 3.3 were extruded with LDPE and characterized for delamination toughness of the multilayered tapes. Normalized delamination toughness of the multilayered film decreased with increasing styrene content. A linear relationship between the delamination toughness and the styrene content in ES copolymers was observed, Figure 3.19. The effect of ES layer thickness on the delamination toughness showed a factor of two increases as the ES layer thickness increased from 10 μm to 15 μm. In low adhesion system with ES60 and ES68 layers, similar to earlier discussed PC/PMMA and PC/SAN25 systems, interfacial failure with bulk stretching of ES layers was observed. In high adhesion system, bulk stretching of ES30 and ES40, and localized stretching of LDPE at the interface confirmed the deformation across the ES layer thickness [32].

3.3.2 Adhesion Properties via Tie Layers

The adhesion properties of different two component multilayered composites demonstrated delamination toughness dependence on the interfacial

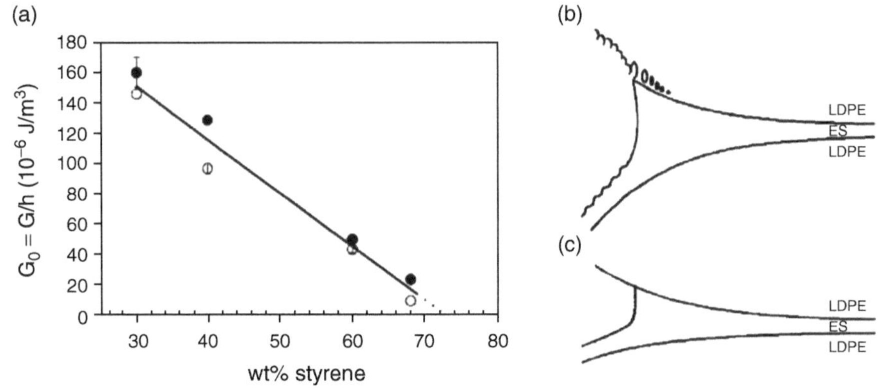

Figure 3.19 (a) Effect of styrene content on delamination toughness normalized to ES layer thickness at ambient temperature and peel rate of 10 mm/min. Closed circles represent 90/10 composition systems, open circle represent 95/5 composition systems; (b) delamination failure mechanism for high adhesion systems LDPE/ES30 and LDPE/ES40; and (c) delamination failure mechanism for low adhesion systems LDPE/ES60 and LDPE/ES68.

adhesion and the relative layer thicknesses two components. Systematic variation of the relative interfacial energies in PC/SAN and LDPE/ES systems was used to control the delamination properties. This approach was further extended to investigate various copolymers as tie layer materials in polyolefinic composites with inherently incompatible components. Use of a third component with its molecular segments compatible with the two components of the composites or blends, known as "compatibilizer" has been practiced. Conventional blending of the polymers with compatibilizer produces variable size domains of the mixed components. Moreover, the blended components are difficult to examine for the interfacial properties due to their three-dimensional (3D) structure. The efficacy of the compatibilizer typically depends upon the blending parameters, efficiency, domain size, etc. Multilayered composites offer unique opportunity to incorporate a compatibilizer resin as a tie layer between the two polymers and controlling the compatibilizer layer (tie layer) thickness down to few tens of nanometers. The composite structure with tie layers is shown in Figure 1.2 in Chapter 1. Similar to PC/SAN, PC/PMMA and LDPE/ES systems, delamination toughness of the layers was used as a tool to investigate the adhesive properties of the compatibilizers.

In this section, effect of compatibilizers on the multilayered films of PP and PE is discussed. PP and PE are the most commonly used polyolefinic materials for their low cost, easy processability and mechanical properties. Incompatibility between PE and PP results from poor interfacial

Table 3.4 Resins Used in Coextrusion of LDPE/ES Multilayered Composites

Resins	Styrene Content (wt%)
ES30	30.3
ES40	39.5
ES60	59.7
ES68	68.4

adhesion, which necessitated the development of compatibilizers such compatibilizers such as – olefinic block copolymers (OBCs), ethylene-octene copolymers (EOs), various elastomers such as EPDM, SBS, and SEBS [33]. The effect of OBC and EO copolymers as tie layers on the adhesion to PP and HDPE was studied to evaluate their use as compatibilizers for blends and multilayered composites. Multilayered tapes consisting of alternating layers of PP and HDPE, separated by OBC and EO as compatibilizer tie layers, were analyzed for delamination properties and damage zone. Tie layer thickness was systematically varied from 15 μm to 100 nm [34,35]. Multilayered composites of PP and HDPE with 65 and 257 layer were fabricated in PP/OBC/HDPE/OBC/PP or PP/EO/HDPE/EO/PP structure. OBC or EO tie layers were present between all PP and HDPE interfaces. The properties of the base materials used for coextrusion are shown in Table 3.4.

The effect of tie layer thickness on the delamination toughness of the composites is shown in Figure 3.20a,b OBC multi-BCPs consistently showed better adhesive properties than the conventional EO copolymers. A linear dependence of delamination toughness on the tie layer thickness was observed at thicknesses greater than 2 μm indicating uniform energy absorption across the entire tie layer thickness. With further reduction in layer thickness, a stronger dependence of delamination toughness on layer thickness was observed. Deformation of entire tie layer thickness with well defined, continuous yielded zones was observed in thicker tie layers, while highly fibrillated and voided craze zone with reduced damage zone length and crack-tip opening were prominent features in layers thinner than 2 μm. It is obvious that the continuous yielding was not supported when the tie layers did not contain enough material for yielding, and, thus, produced voids. In both EO and OBC tie-layer systems, delamination occurred preferentially at PP/tie-layer interface as revealed by the FTIR analysis of delaminated surfaces. In general, OBC and EO copolymers exhibited better adhesion to HDPE layers than PP layers. Schematic representation of the tie-layer damage zone is shown in Figure 3.20c and d.

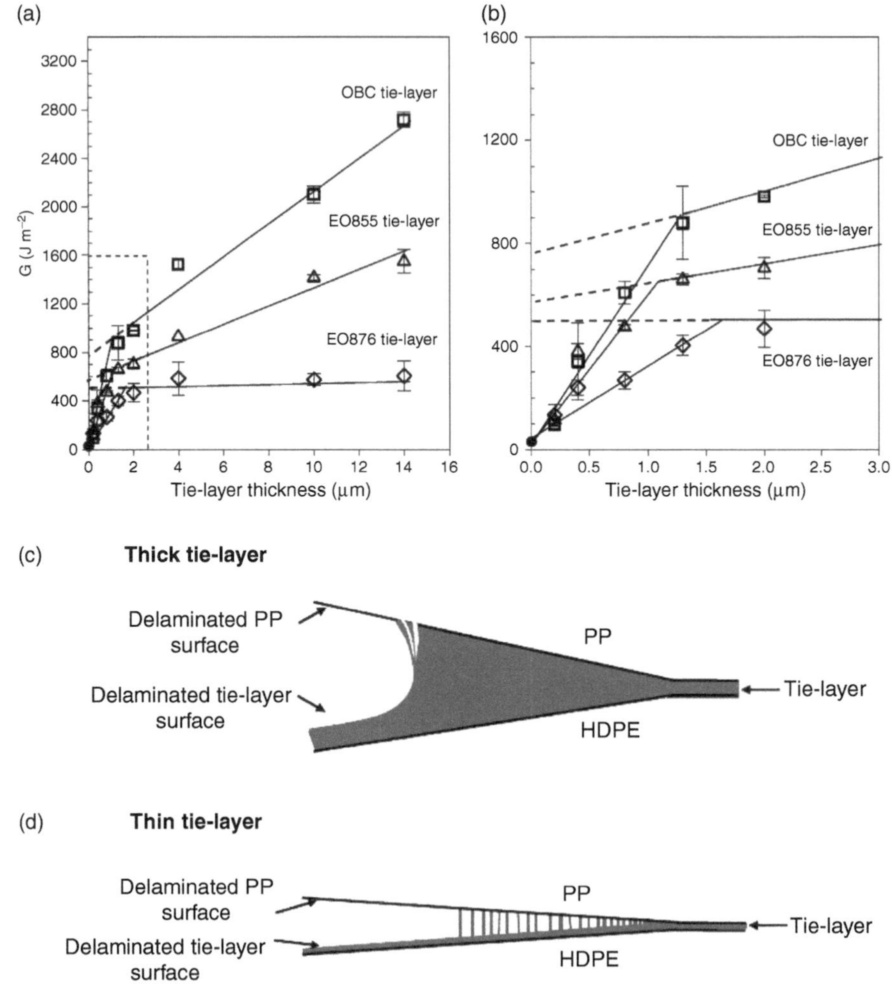

Figure 3.20 Effect of tie-layer thickness on the delamination toughness. (a) The entire range of tie-layer thicknesses studied and (b) a magnified plot of the results for thinner tie layers. The cross-head speed was 10 mm/min and the temperature was 21°C; schematic representation of the tie-layer damage zone. (c) Thick tie layers (2–14 μm) and (d) thin tie layers (less than 2 μm).

Maximum delamination toughness was observed close to the T_g of the tie layer materials. Reduction in the test temperature resulted into catastrophic failure due to absence of stable crack propagation below the T_g. In general, with increasing temperature above the T_g, the delamination toughness decreased due to changes in the modulus and yield stress, which negatively affected the delamination [34].

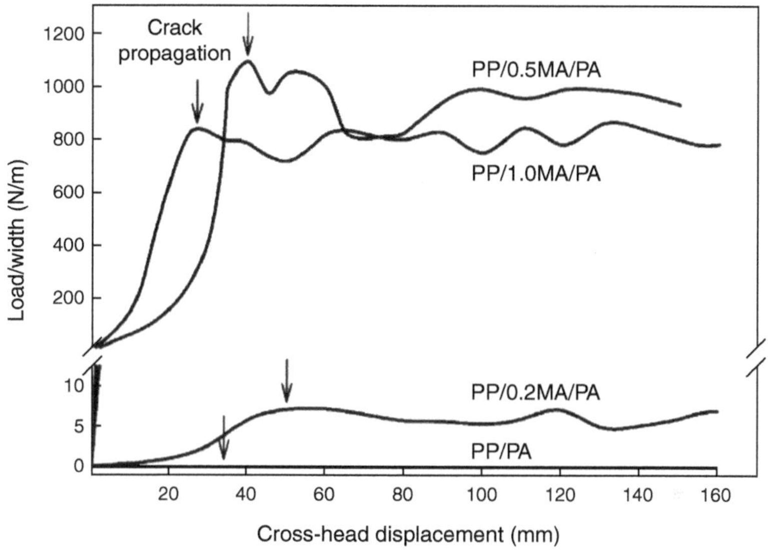

Figure 3.21 Schematic representation of the tie-layer damage zone. (a) Thick tie layers (2–14 μm) and (b) thin tie layers (less than 2 μm).

In another example, with PP-*g*-MA as a tie layer for PP and polyamide 66 (PA66) multilayered films, increasing the maleic anhydride content of the tie layer improved the interfacial toughness of the layers as shown in Figure 3.21 [36]. Interfacial delamination during delamination test was obvious due to poor adhesion between PP and PA6. When PP-*g*-MA was used as a tie layer, measured peel toughness of 970 ± 170 J/m^2 was two orders of magnitude higher than the samples without the tie layers. This change in toughness was attributed to change in the peel mechanism from interfacial delamination to craze delamination [36].

Expansion of the delamination studies of polymer interfaces using coextrusion was to understand the effect of chain architecture on the adhesion of ethylene copolymers with PP. Combining the two most widely used polymers for improved properties has led to the development of wide range of ethylene copolymers. Heterogeneous Ziegler–Natta catalyzed ethylene copolymer (ZNPE), metallocene catalyzed PE (mPE) and their blends were coextruded with PP [37,38]. The effect of mPE content on the adhesion of ZNPE blends to PP is shown in Figure 3.22a. Poor adhesion between ZNPE and PP due to segregation of low molecular weight, highly branched fractions in ZNPE at the interface creating an amorphous interfacial layer was unable to form entanglement bridges with PP. Conventional Ziegler–Natta catalysts typically produce wide molecular weight distribution with significant low molecular weight chain fraction. Interestingly, the blending of ZNPE with mPE resulted into epitaxial crystallization of ZNPE chains on PP thereby increasing the delamination toughness. Increasing mPE content in the blends increased the delamination toughness in the layered composites linearly. Interfacial regions with epitaxial crystallization of ZNPE bulk chains and mPE chain entanglement contributed to improve the overall delamination toughness. A detailed model for interfaces of ZNPE, mPE and their blends is shown in Figure 3.22b [38,37].

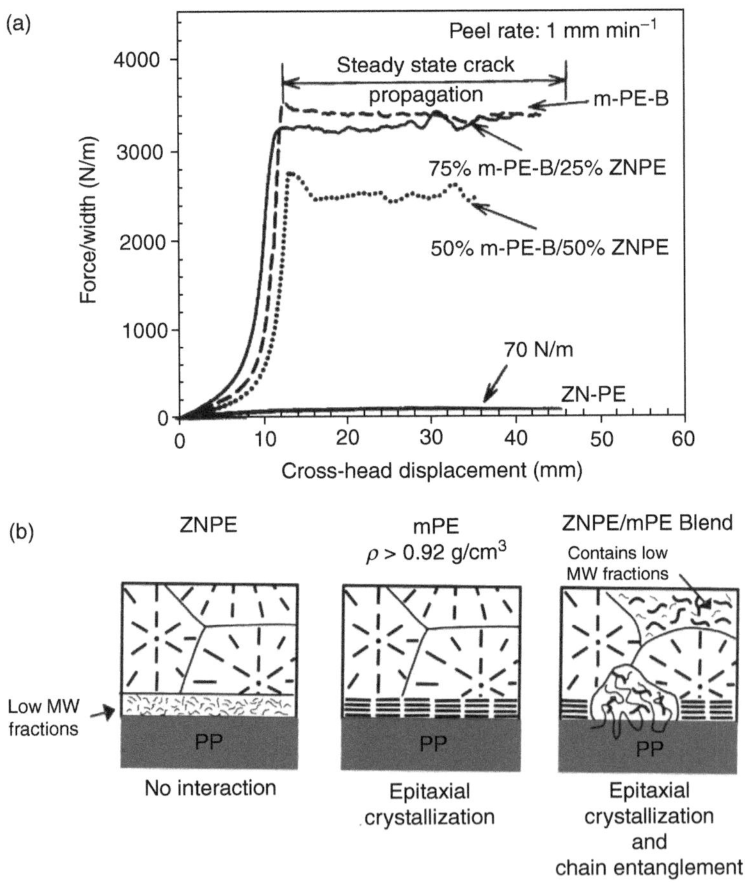

Figure 3.22 (a) Effect of metallocene PE content on adhesion of polyethylene blends to PP and (b) model for PP and ZNPE, mPE, and blends interfaces.

3.3.3 Mechanical Properties of Multilayered Composites

Mechanical behavior of multilayered composites was investigated in a series of multilayered composites of ductile PC and brittle SAN polymers; with varying relative concentrations of the individual components [39]. The effect of the relative composition of the two components on the tensile properties of 49-layered composite, measured at 10%/min strain rate, is shown in Figure 3.23a. Brittle fracture in SAN control at low strain

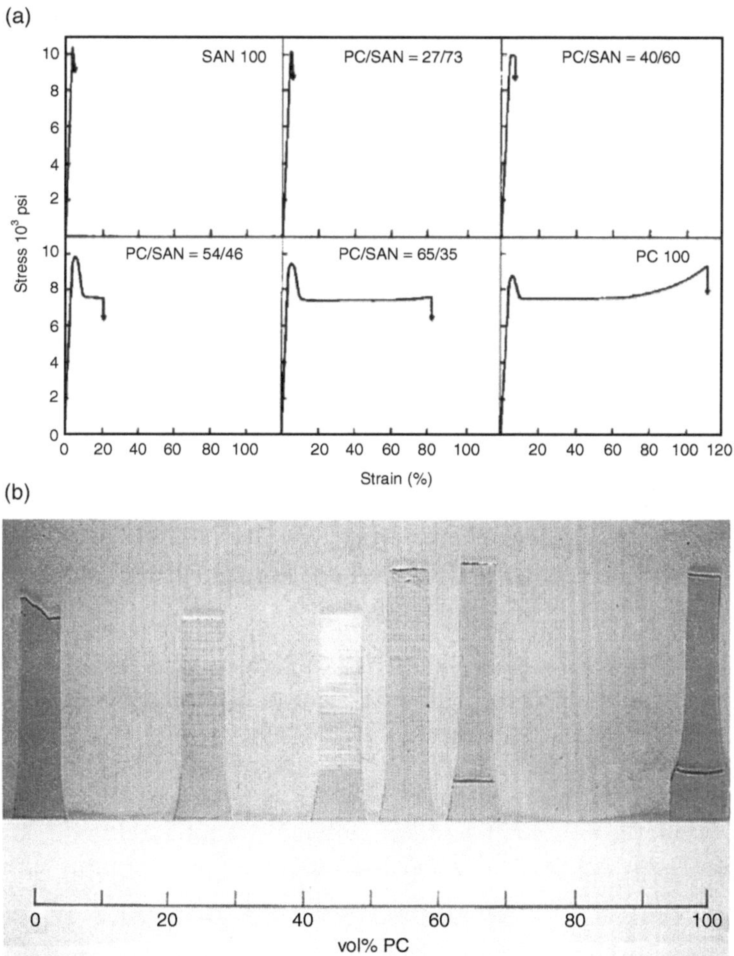

Figure 3.23 (a) Stress–strain curves of 49-layered PC/SAN tensile specimens with variable relative composition and (b) tensile specimens from 49-layered PC/SAN composites showing changes in fracture mode with % PC.

of 2.5% and cold drawing of PC with tensile failure close to 120% was measured. Intermediate compositions showed two types of stress–strain behavior. Samples with SAN as a major component showed fracture prior to yielding. As the SAN fraction decreased below 50%, fracture stress drop post yielding, significant ductility and increased elongation at break were observed with increasing PC fraction.

The fractured samples are shown in Figure 3.23b [40]. The parallel cracks across the width of the samples were observed in composites with higher SAN content, 27/73 and 40/60 PC/SAN compositions. Other samples, 54/46 and 65/35 PC/SAN composites showed reduced number of cracks and the presence of shear bands, which propagate along the length of the sample, deforming the PC and SAN layers. The brittle to ductile transition was, therefore, observed in samples with 54% or more PC in the measured compositions. The multilayered composites also showed specimen size dependent deformation behavior. When the sample width changed to 0.5 mm, a 65/35 sample crazed excessively as compared to cold drawing in standard tensile sample with 1.27 cm width (standard tensile specimen of ½"). It is suggested that the free edge effects in the wider sample initiate cracks from the middle of the sample and stop prior to reaching the free edges, while crazes can easily initiate from the specimen edges in smaller width samples. Brittle to ductile transition shifted to lower PC content of 45% in the composites with increased number of layers to 193. It is obvious that the number of layers and the relative composition of the two components can impact brittle to ductile transition even further [39].

3.3.4 Layer Thickness Effect on Mechanical Properties

Effect of layer thicknesses on the mechanical properties was investigated in multilayered composites of PC/SAN25 with 49, 194, and 776 layers fabricated with the volume ratio of the two components as 65/35 [41]. An additional composite with 388 layers and 50/50 composition was also fabricated. The overall thickness for all the composites was maintained at 1.2 mm. The stress–strain behavior of 49-, 194-, 388-, and 776-layered composites, measured at 54%/min strain rate, is shown in Figure 3.24. All four samples showed ductile deformation with necking, and comparable modulus of 2.6 GPa and yield stress of 70 MPa. The modulus of the composites followed rule of mixtures and the yield stress was intermediate between yield stress of PC (62 MPa) and SAN25 (82 MPa). However, the fracture strain increased with increasing number of layers, which was associated with changes in the microdeformation behavior. In all samples, primary crazes formed in SAN layers initiated

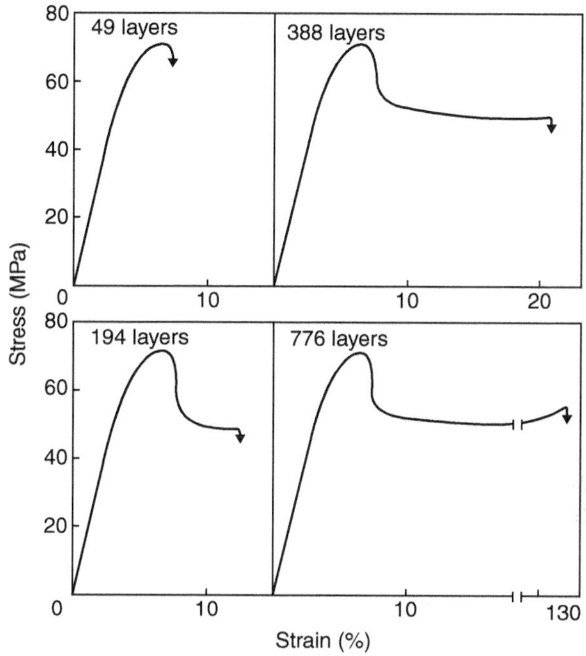

Figure 3.24 Stress–strain curves of 49-, 194-, 388-, and 776-layered PC/SAN composites.

at the interface, and subsequent higher stresses resulted in shear deformation in PC layers. Deformation of the 49-layered films showed formation of crazes in SAN layers, which initiated at the interface followed by propagation across the width of the SAN layer thickness. Increased strain resulted into secondary crazes in SAN layers and localized shear deformation in PC layers, which initiated from the primary SAN crazes at the PC/SAN interfaces. Shear zones across the entire PC layer occurred only at higher strains close to yielding. When the strain is high enough, close to yielding, SAN crazes opened into cracks and shear banding resulted into drawing of the PC layers. The two phenomena showed reduced PC layer thicknesses and void opening in SAN layers without layer thickness reduction leading to eventual sample failure. Detailed features of the microdeformation of 49-layered composite are shown in Figure 3.25a–d. A similar failure mechanism was observed in 194-layered composites. The 49-layered composites showed immediate fracture prior to necking, while 194-layered composites showed some neck propagation prior to fracture at 12% strain. In these composites, SAN crazing and PC shear deformation, characteristic of bulk materials, were present.

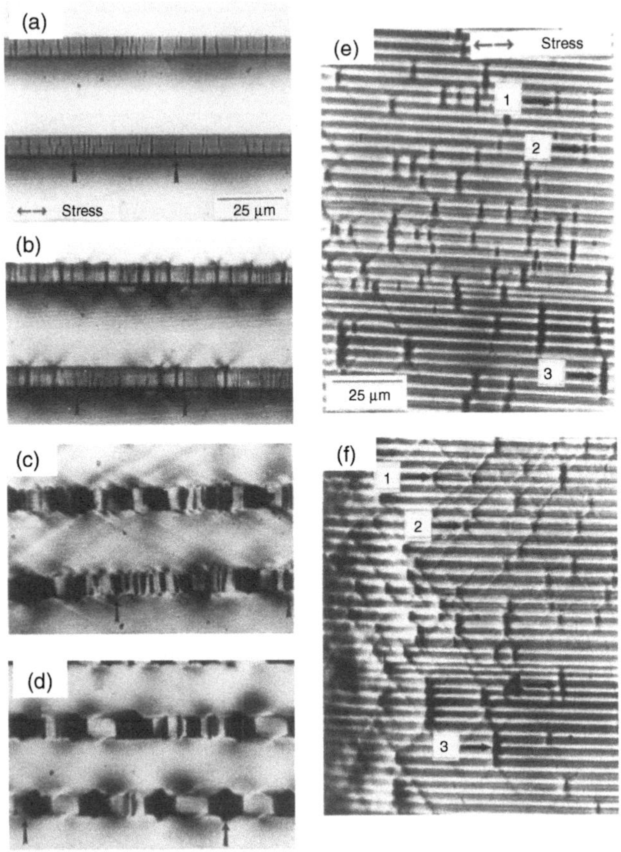

Figure 3.25 Optical micrographs of microdeformation of (a–d) 49-layered composite with increasing strain. The arrows on each micrograph mark the same region of the specimen. (e,f) Details of the necking region of 388-layered composite showing the progression of necking.

In 388 and 776-layered composites, crazing suppression and formation of the shear bands followed by extension through several PC and SAN layers changed the mechanical properties significantly, as shown in Figure 3.25e,f. Neck propagation across the entire gauge length was observed in 776-layered composites. The deformation in these samples showed crazing and cracking of SAN layers, similar to 49- and 194-layered composites. However, shear bands initiated at crack tip extended through multiple SAN and PC layers. As the yielding process occurred, SAN layers are drawn along with PC layers in ductile manner, changing the SAN deformation mechanism from brittle fracture to shear yielding [41].

3.3.5 Deformation in Multilayered Films

Further investigation of crazing phenomena in PC/SAN composites was performed by changing the relative thicknesses of the PC and SAN components as well as the number of layers [42]. Similar to earlier work, 49-, 194-, 338-, and 776-layered composites, and additional 927- and 1857-layered composite were produced. The relative volume composition of PC/SAN was varied between 28/72 and 76/24 to achieve layer thicknesses ranging from 400 nm to 30 μm. Although the overall deformation behavior was similar to 65/35 PC/SAN composites discussed earlier, three types of crazing were observed with changes in the relative thicknesses of PC and SAN layers: (a) randomly distributed single crazes in SAN layers, (b) craze doublets with aligned crazes in neighboring SAN layers, and (c) craze arrays with many aligned crazes in neighboring SAN layers. Examples of these mechanisms and summary of results are shown in Figure 3.26. The crazing behavior was strongly dependent on the PC layer thickness in all samples. When PC layer thickness was greater than 6 μm, formation of single crazes was predominant. As the layer thickness decreased, formation of craze doublets between 1.3 μm and 6 μm was observed. Further reduction in PC layer thickness produced craze arrays [42].

Deformation behavior of PC/PMMA multilayered composite with better adhesion properties than the PC/SAN composites showed significant differences in the yielding behavior [43]. For example, a 32-layered PC/PMMA composite (70/30 composition) showed a sharp stress drop indicating yielding as shown in Figure 3.27. The yield maximum on the PC/PMMA stress–strain curve broadened with increasing PMMA content. Such broadening of the yield was not observed in PC/SAN composites. All PC/PMMA composites showed ductile failure irrespective of the PMMA content, unlike brittle failure in PC/SAN composites with high SAN content. Microdeformation behavior of the composites showed crazing of PMMA layers at low strains, followed by initiation and growth of the microshearbands at the craze tip at the PC/PMMA interface with further strain increase. The microshearbands from the opposing interfaces in PC layers coalesced with increasing strain at yield point. Some of the PMMA crazes formed microcracks tearing into adjacent PC layers. As the interfacial adhesion was significantly stronger in PC/PMMA system, no delamination failure was observed in these composites. Interestingly, when the samples were deformed at 10%/min strain rate, yielding of the PMMA layers with improved yield stress was observed. Some crazes also opened up at the center of the PMMA, further supporting observed strong interfacial adhesion, Figure 3.28b. In comparison, 70/30 PC/SAN sample tested at 0.1% strains showed local delamination at the interface, Figure 3.28c.

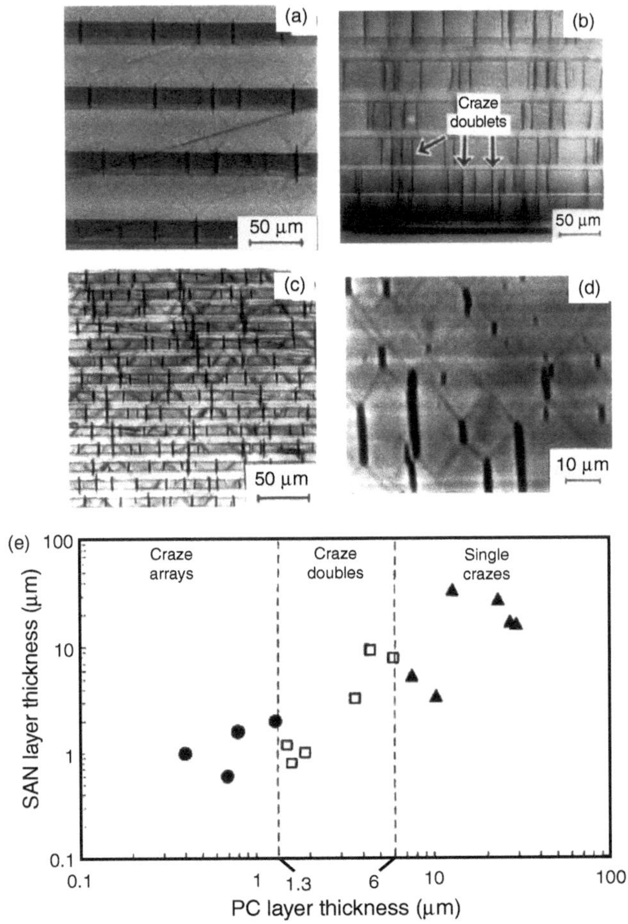

Figure 3.26 Optical micrographs. Of (a) 49-layered PC/SAN (26/16 μm) at 6% strain; (b) PC/SAN (13/33 μm) 49-layered specimen edge region showing decreasing PC layer thickness with subsequent registration in neighboring SAN layers; (c) 194-layered specimen with PC/SAN (5.9/7.9 μm) with craze doublets, microshearbands at 6% strain before necking; (d) 927-layered specimen with PC/SAN (0.8/1.6 μm) with craze doublets, microshearbands at 6% strain before necking; and (e) type of crazing as a function of PC and SAN layer thickness.

Prior to yielding, both systems showed similar microdeformation behavior where crazing in PMMA and SAN layers caused microshearband and the subsequent shearband coalescence at higher strains in PC layers. The formation of microcracks in SAN layer was also similar to occasional microcrack formation in PMMA layers. However, the PMMA cracks translated into PC layer unlike PC/SAN composites, while delamination at PC/SAN interface occurred to relieve the SAN layer constrains permitting

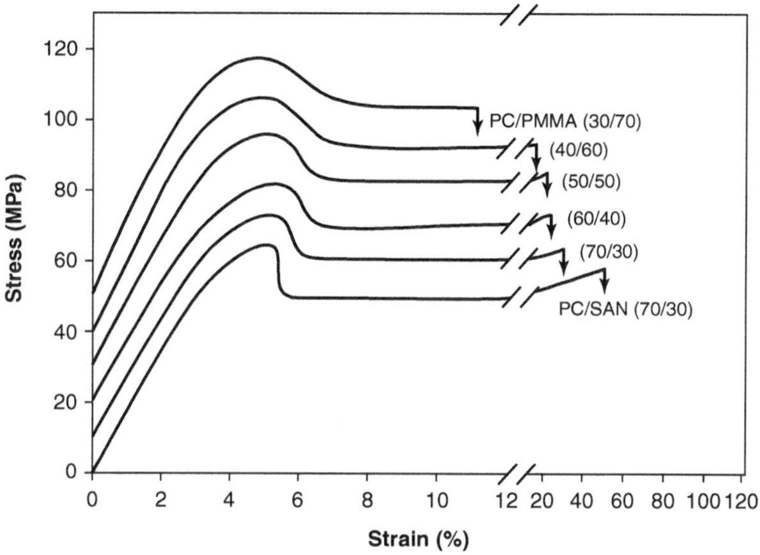

Figure 3.27 Stress–strain curves at 0.1%/min for 32-layered PC/PMMA specimens of various compositions, curves are shifted vertically by 10 MPa. A PC/SAN (70/30) specimen is included for comparison.

drawing out of PC layers. Multilayered composites of PC/PMMA system with increased number of layers showed improved ductility. The stress–strain curves of 32-, 256-, and 2048-layered PC/PMMA (70/30) composites tested at 50%/min strain rate show increased fracture strain from 4% in 32-layered samples to 30% in 256 and 60% in 2048-layered samples as shown in Figure 3.29 [43].

Investigation of multilayered composites with two ductile components was studied in PC/PET composites. Multilayered composites of PC and PET produced with varying number of layers from 8 to 1024. The stress–strain properties of the 50/50 PC/PET specimens (layer thicknesses ranging from 125 μm down to 900 nm) showed stress maximum, cold drawing and necking behavior; which did not change with layer thickness

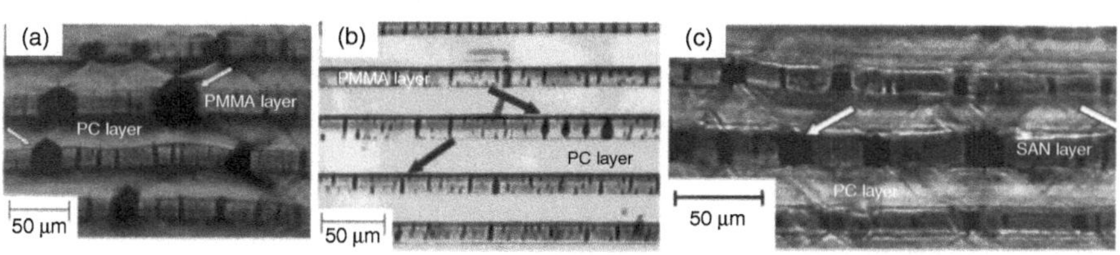

Figure 3.28 Optical micrographs of edge view. Of (a) 32-layered PC/PMMA (70/30) tested at 0.1%/min; (b) 32-layered PC/PMMA (70/30) tested at 10%/min; and (c) 32-layered PC/SAN (70/30) tested at 0.1%/min.

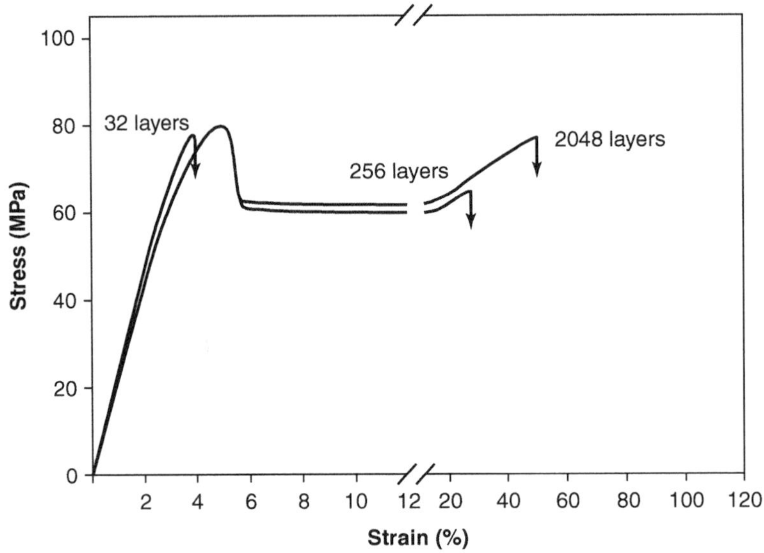

Figure 3.29 Stress–strain curves at 50%/min for PC/PMMA (70/30) with 32, 256, and 2048 layers.

or the number of layers. However, the deformation behavior was significantly different than the PC/SAN and PC/PMMA systems. The composite structures with individual layer thickness less than 1 μm exhibited shear zones across the entire sample thickness resembling a single component behavior and samples with thicknesses greater than1 μm, showed localized shear bands in individual layers as shown in Figure 3.30 [44,45]. The mechanical properties such as Young's modulus, yield stress and stress at break of the multilayered composites were the average of the individual components. Elongation at break of 150% was significantly less than the additive average of PC and PET elongation at break of 100 and 400%,

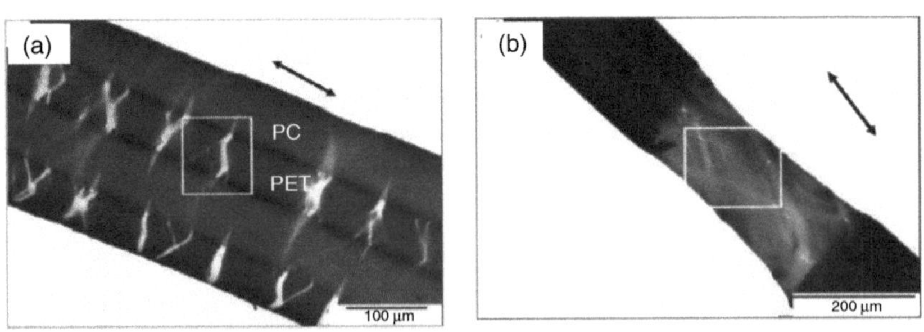

Figure 3.30 TEM micrographs of deformed thin sections. Of (a) eight-layered PET/PC (50/50) samples showing shear bands in PET phase and (b) 1024-layered PET/PC (50/50) sample showing shear bands through the whole film.

respectively. The thermal annealing of the composites at 135°C for 72 h impacted the mechanical properties of the control samples; however, did not affect the mechanical behavior of the layered composites. Typically, PET crystallization during annealing leads to its embrittlement as evident in monolayer control samples, which showed an elongation at break of only 50%. On the other hand, the physical aging of PC at the annealing temperatures showed increased yield stress indicative of reduced ductility. Overall improvement in the yield stress in layered composites without any negative impact on the strain at break was indeed due to layering. Analysis of composite samples with individual layer thicknesses below 100 nm also showed homogeneous plastic deformation [44].

3.3.6 Impact Properties of the PC/SAN Multilayer Composites

Impact strength of the PC/SAN composites were measured in notched samples and normalized with respect to 100% PC control (214 J/cm) to obtain the relative impact strength shown in Figure 3.31 [39]. The impact strength increased with increasing PC content for a given number of layers, as expected. Interestingly, with increasing number of layers,

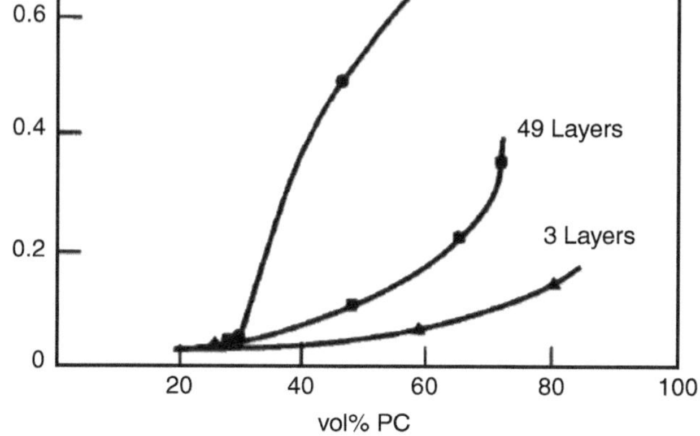

Figure 3.31 Relative rheometrics impact strength (at room temperature) of PC/SAN composites, with number of layers and PC content as two variables (relative impact strength of PC = 1.0 and SAN = 0.0).

the impact strength increased significantly for the same PC content. In a 391-layered samples, the impact strength of 55/45 PC/SAN composite is 95% of the impact strength of the 100% PC. In addition, the brittle to ductile transition, indicated by sharp increase in impact strength, showed a steady shift to lower PC content with increasing number of layers. For example, in 49-layered composite, the brittle-ductile transition required 65–75% PC, while a sharp rise in the impact strength for 193-layered composites was observed in samples with less than 40% PC content. The increased toughness of the multilayered composites with increasing number of layers (i.e., decreasing layer thickness) was attributed to the presence of number of crazes and cracks in SAN layers, which initiate the microshearbands in PC layers. The shear banding in PC layers act as a toughening mechanism. It is intuitive that more number of crazes would initiate large number of microshearbands in PC layers, increase the energy absorption and result into toughness improvement. Similar improvement in the impact properties are also observed in PC/PMMA multilayered sheets [46]. To investigate the number of crazes as a function of layer thickness, a multilayered composite system of a general purpose PS and styrene/butadiene multilayered composite structure with wide range of layer thicknesses within the same sample was produced. The deformed samples showed more crazes in thinner layers than the thicker layers in this sample. Therefore, the number of crazes in a composite with thinner layers (more number of layers) is more than the composites with thick layers (less number of layers). The cracks also develop shear bands in the adjacent layers, which are energy absorption sites. It is evident that multilayered composites of PC/SAN with increased number layers showed improved impact properties due to increased sites for energy absorption, even when the composition maintained constant. With new capabilities to produce gradient layer thicknesses, it is possible to manipulate the toughening of the multilayered composites and achieve different failure phenomena within the same composite. An example of such modification is discussed in the later part of this section.

3.3.7 Fatigue in Multilayered Films

The effect of layering on fatigue crack propagation (FCP) was investigated in PC/SAN composites. The crack propagation rates ($\Delta L/\Delta N$, mm/cycle) as a function of energy release in 49- and 776-multilayered composites of PC/SAN 65/35 are shown in Figure 3.32a [47]. The fatigue resistance is measured by critical strain energy release rate J_{1C} at which the spontaneous crack propagation occurs. A significant different in J_{1C} values from 8 kJ/m^2 to 34 kJ/m^2 was observed when the number of layers

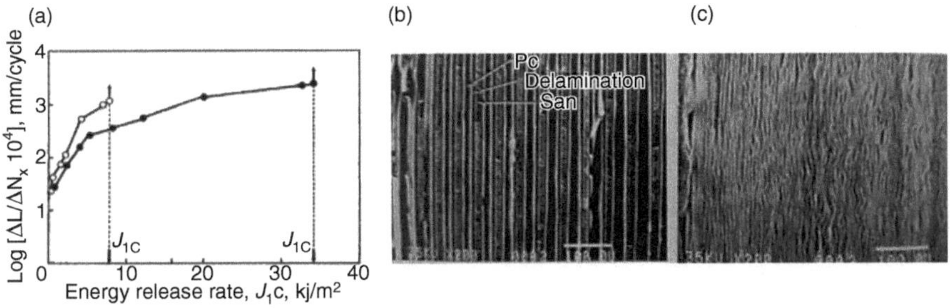

Figure 3.32 (a) Strain energy release rate of the 49 (o) and 776 (●) layer composites versus crack growth per fatigue cycle during stable crack propagation; (b) SEM image of fatigue fracture surface of 49-layered composite; and (c) SEM image of fatigue fracture surface of 776-layered composite.

increased from 49 to 776. Ability to absorb more energy per volume of the damage with higher number of layers increases the fatigue resistance in layered composites. 776-Layered composites also showed increased number of fatigue cycles from crack initiation to failure and a longer stable crack length. Failure features of the 776-layered composite showed a ductile features while 49-layered composite showed catastrophic failure. The ability to draw SAN in thinner layers along with PC contributes to ductile features as shown in Figure 3.32b,c [40]. A significant advantage is envisioned in using polymer composites in high-pressure applications such as pipes [40,47].

3.3.8 Filled Multilayered Composites

Multilayered films with inorganic fillers in one of the layers have shown significant improvement in the gas barrier properties. Similar enhancements in the mechanical properties in the filled multilayer polymer systems were demonstrated with talc as filler. Coextrusion of talc filled PP with unfilled PP produced 50/50 (vol./vol.) composites with the number of layers varying from 16 to 1024 [48]. Talc particles with reported thickness of 1 μm and aspect ratio of 10 were blended with PP at 40 wt% concentration prior to coextrusion processing. Multilayered sheets showed distinct talc-filled PP and unfilled PP layers in composites with 256 or fewer layers. The PP layer thickness in these samples ranged from 60 μm in 16-layered composites to 4 μm in 256-layered composites. It should be noted that the filler thickness in these composites is sufficiently less than its PP layer matrix. As the layer thicknesses became comparable with talc filler thickness, it was difficult to distinguish between the alternating layers of filled and unfilled PP in 512- and

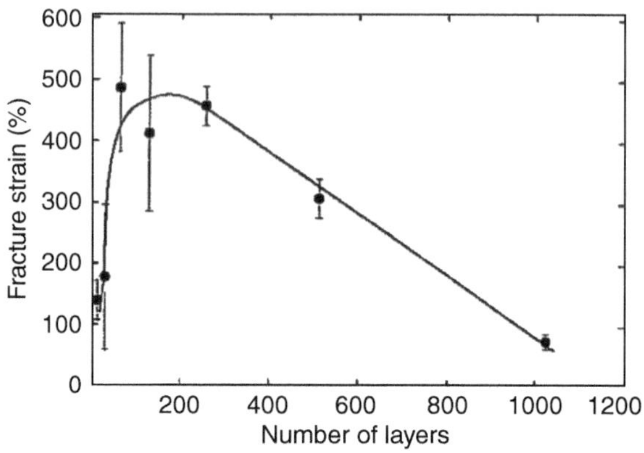

Figure 3.33 Effect of number of layers on fracture strain of multilayers of PP and 40% talc-filled PP.

1024-layered composites. Mechanical properties measurements showed gradual increase in fracture strain as the number of layers changed from 16 to 64 layers with a property maximum between 64 and 256 layers as shown in Figure 3.33. Further increase in the number of layers resulted into quasibrittle failure with reduced fracture strain.

Morphology analysis of composites with 16- and 32-layered composites showed poor talc-filler orientation, which results into higher probability of crack initiation. Once the crack reaches a critical size, unfilled PP layers are unable to stop crack propagation. In thinner layers with layer thicknesses comparable with filler thickness, the filler particle agglomeration acted as stress concentrators accelerating crack initiation significantly. This effect was minimized in the composites with layer thicknesses in 4–15 μm range. The results can be used as a reference to identify critical layer thicknesses for improving fracture properties in filled systems. Further investigation on offset polymer compositions can help understand the fracture mechanism in filled system [48].

3.3.9 Gradient Layer Composites

Multilayer coextrusion process flexibility to create gradient structures is discussed in Chapter 2. Gradient multilayered films of styrene/saturated ethylene–propylene block copolymer (PS/SEPS) were produced using a series of "uneven" split multipliers. A 256-layered film with a 10× gradient in the layer thickness were produced with layer thickness change from 1 μm to 100 nm within the same film. The gradient composite was deformed to a strain of 2% to show crazing in thicker layers, while thinner layers showed long shear bands along with short crazes on the film

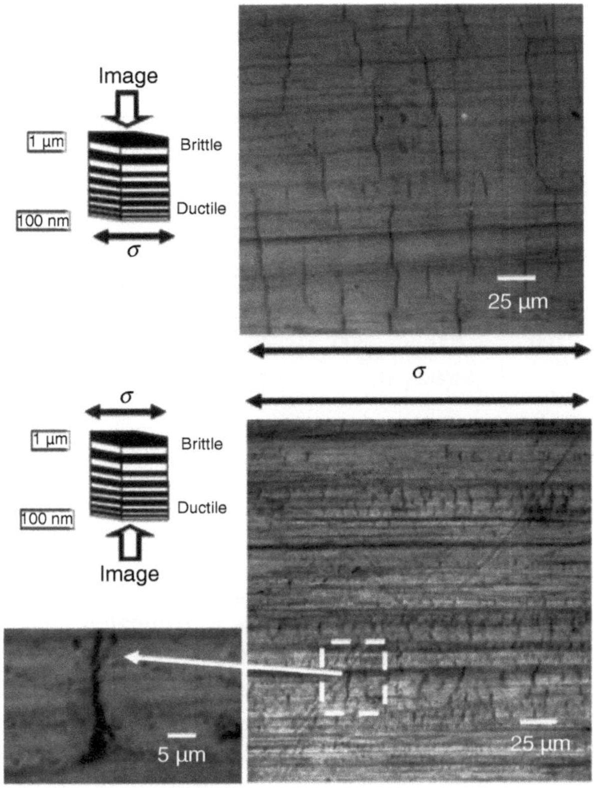

Figure 3.34 Optical microscopy of 256-gradient layered 50/50 (vol./vol.) PS/SEPS film. (Top) film surface micrograph of thicker layers exhibiting crazes at 90° to the film deformation direction. (Bottom) film surface micrograph of thinner layers exhibiting large shear bands and smaller crazes that terminate into shear bands as a result of deformation.

surface, Figure 3.34. The PS deformation transition from crazing in thick layers to cooperative crazing to shear banding in thin layers was indicative of increased PS and SEPS interaction across the composite thickness [49]. The ability to control the gradient structures and to produce multilayered structures with variable layer thickness distributions allows designing of complex systems [49].

3.3.10 Nanoscale Confinement Effect on Mechanical Properties

Confinement of semicrystalline polymers such as PEO, PCL limits the spherulitic crystallization in nanolayered multilayered films as discussed

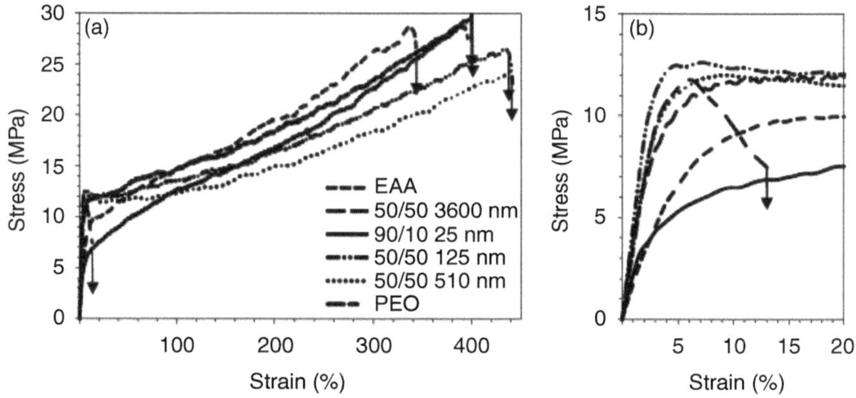

Figure 3.35 Stress–strain curves of EAA/PEO and controls (room temperature, 100%/min). Showing (a) films retained ductility of EAA and (b) increase in the modulus and yield stress, and decrease in the yield strain with decreasing layer thickness.

in Section 3.1. Coextruded films of PEO against EAA or PS systematically changed the morphology from spherulites to truncated spherulites or discoids, to oriented in-plane lamellar structure with decreasing layer thickness from micro- to nanoscale. At thicknesses close to 20 nm, the confined morphology of PEO layers showed single lamellae with large aspect ratio, resembling single crystal structures. The effect of confined structures was investigated and modeled to identify the contribution of crystalline morphology and amorphous phase on mechanical properties. PEO/EAA multilayered films with 33, 257, and 1025 alternating layers with 50/50 (vol./vol.) composition were produced with overall film thicknesses ranging from 50 μm to 130 μm. The individual PEO layer thickness changed from 45 nm to 3600 nm [50,51]. A 1025-layered film with 10/90 PEO/EAA composition with 25 nm PEO layer thickness was also produced. Stress–strain behavior of the films was measured in uniaxial tension as shown in Figure 3.35. PEO control films exhibited brittle fracture at 14% and EAA control showed ductile behavior with fracture strain of 340%. Since the confinement did not impact the EAA crystallization behavior, the modulus change in EAA layers was assumed to remain constant. The calculated PEO modulus shows a threefold increase in the modulus as compared to PEO control. With decreasing layer thickness from 3600 nm to 45 nm, the tensile modulus increased from 486 ± 84 MPa to 1450 ± 99 MPa, at room temperature measured at 100%/min strain rate. Low temperature measurements at −10°C at 5%/min strain rate also showed a similar change as the modulus increased from 730 ± 80 MPa to 1240 ± 50 MPa over the same thickness range. PEO layers offered a unique opportunity to change the lamellar orientation from in-plane to on-edge by melt recrystallization approach, with selective melting of PEO layers followed by fast quenching, as described in Section 3.1. The measured tensile modulus in on-edge lamellae was independent of PEO layer thickness. PEO/EAA model system demonstrated a significant impact of the lamellar orientation on the modulus of multilayered film [50,51].

Deformation mechanism of the layered composites was investigated by stretching the films to different strains, from 0% to 400% and analyzing the structure of PEO/EAA layers with respect to different crystal populations in PEO layers. In-plane, on-edge, and mixed orientations were observed in the films at different strains. In 366 and 510 nm PEO layers, isotropic orientation of PEO layers in extruded films changed to on-edge lamellae with increasing strain of up to 100%. With further increase in strain to 400%, PEO chains became aligned with the deformation axis by chain pull-out and recrystallization mechanism. More than 85% orientation of the chains along the deformation axis was similar to PEO-fiber structure. As the layer thickness decreased to 125 nm, the spherulitic morphology in thick layers changed to loosely aligned stacked lamellae, in the layer direction in extruded films. Further reduction in PEO layer thickness to 25 nm showed single PEO lamellae, resembling single-crystal structure. Stretching of the films showed nonuniform deformation (micronecking) and lamellar alignment in the deformation direction. 25 nm layers showed strain-induced crystallization with 40% crystals still remained in-plane. The large amount of in-plane crystals maintained in the 125 and 25 nm layers was different than the thick layer deformation, where almost complete orientation of chains to deformation axis was observed. AFM images of deformation behavior in PEO layers are shown in Figure 3.36. Similar to PEO, the crystal orientation changed significantly in nanolayer films of PCL, PE, sPP, and iPP as demonstrated earlier. Possibility of similar shifts in the mechanical properties of multilayer composites containing these polymers has been suggested [51].

3.3.11 Mechanical Properties of Confined Block Copolymers

Multilayer coextrusion approach was extended to confine the BCPs to investigate the effect of the BCP microstructure on the mechanical toughness. Such 257-layered films of polystyrene-block-polyethylene/polypropylene-block-polystyrene triblock copolymer (SEPS) copolymer and PS, were investigated [52,53]. The SEPS/PS composition was maintained as 50/50 (vol./vol.) with the overall film thicknesses ranged from 25 μm to 250 μm. The individual layer thicknesses varied between

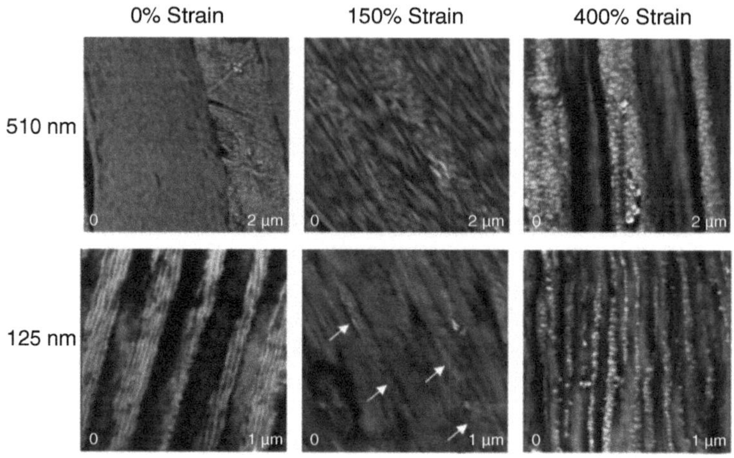

Figure 3.36 AFM images of PEO layer structure in PEO/EAA multilayered films with increasing strain. (Top row) 510 nm PEO layers; (bottom row) 125 nm PEO layers. The layer direction is vertical in the images.

620 nm and 100 nm. Uniaxial tensile deformation of PS film showed brittle fracture at <2% strain, while SEPS film showed longer elongation at break of 490%. Films with layer thicknesses 620 and 480 nm showed brittle fracture at strains less than 5% with the formation of crazes propagating at 90° to applied stress. Further reduction in layer thickness to 190 nm resulted in a higher strain value and showed shear banding with striations at 45° to the applied force. This brittle to ductile transition was noted with decreasing layer thickness in SEPS. An interesting effect of layer thickness was observed on PS cylinders orientation and their hexagonal packing. In the films with layer thicknesses greater than 190 nm, hexagonal packing of PS cylinders in SEPS layers and their isotropic orientation was confirmed. With decreasing layer thickness, PS cylinders were present parallel to the layer direction packed uniformly across the layer thickness. The highly oriented PS cylinders in the SEPS layers resulted in uniform deformation mode in the layered films improving the toughness [52] (Figure 3.37).

3.4 Interphase Materials

In polymer composites, an interphase region is defined when two polymers with different interaction parameters are brought together resulting in a localized mixing of polymer chains at the polymer–polymer domain boundary (interface). Compatibility of polymer blends, adhesion of blend

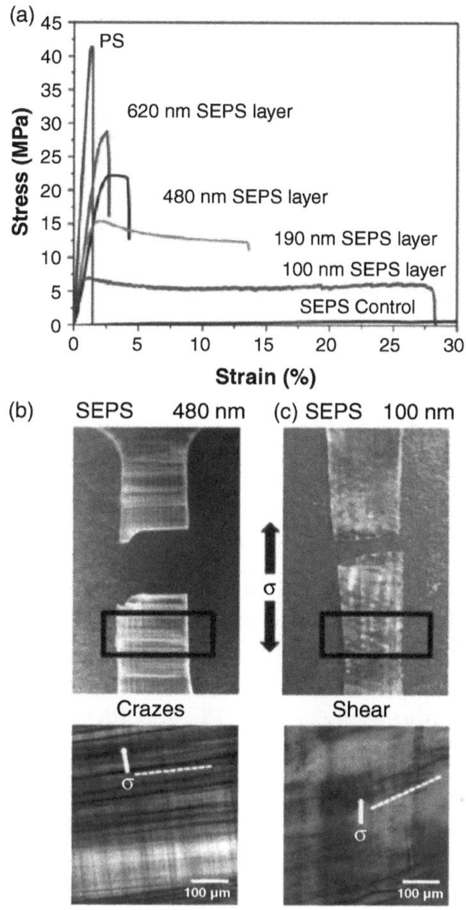

Figure 3.37 (a) Stress–strain response of PS/SEPS multilayer films under uniaxial deformation. Optical micrographs of deformed PS/SEPS multilayer films; (b) 480 nm SEPS layer thickness; (c) 100 nm SEPS layer thickness.

components, other properties such as thermal, mechanical, gas transport, and optical are dependent on the interphase properties. Helfand and coworkers developed a generalized theory to define interphase thickness dependence on the thermodynamic interaction parameter [54]. Because the typical interphase thicknesses are estimated in the range of 5–10 nm, it was difficult to measure the interphase thicknesses experimentally. Additionally, interphase fractions in the blends are extremely small, making it difficult to characterize their properties and its impact on the physical properties of polymers. Multilayer coextrusion processing allows processing of films with thousands of layers with individual layer thicknesses down to few

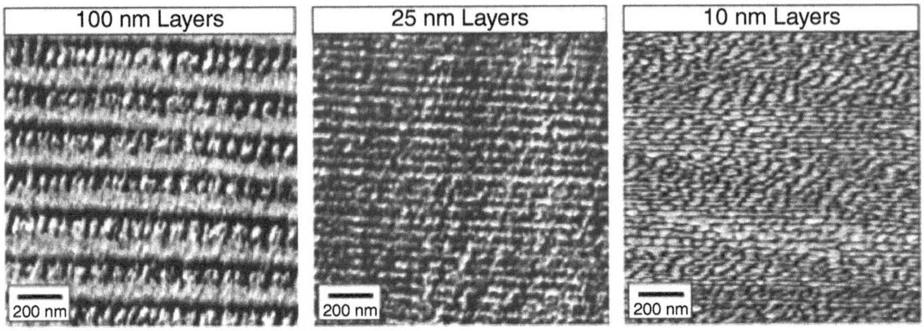

Figure 3.38 Atomic force microscopy (AFM) phase images of microtomed cross-sections of PC/PMMA nanolayer films. Films with average layer thickness 100 and 25 nm show continuous, uniform layers of PC (smooth and bright), and PMMA (hackled). The film with an average 10-nm layered thickness shows an irregular periodicity, which is about twice the average layer thickness.

nanometers. By producing films with the layer thicknesses comparable to the interphase thicknesses, thousands of interphases were successfully created in the multilayered film systems. As the layer thicknesses decreased, the amount of interphase in the composite increased significantly allowing its analysis using conventional characterization techniques.

An immiscible polymer system of PC and PMMA was selected to create and characterize interphase materials [55]. A multilayered film of PC/PMMA with 50/50 composition was extruded with increasing number of layers and varying film thicknesses from 200 μm to 20 μm. Calculated layer thicknesses ranged from 25 μm to 5 nm. As revealed in the cross-section AFM images of the films in Figure 3.38, continuous layers of PC/PMMA were observed down to 25 nm thickness while some layer thickness nonuniformity was noticed in 10 nm layers. Thermal characterization of the PC/PMMA films revealed interesting phenomena of T_g temperature shift with decreasing layer thicknesses. Measured T_gs of PMMA and PC controls at 112 and 144°C were clearly observed in layer thickness 1 μm and slight deviations were noted in layer thicknesses greater than 100 nm. Below 100 nm layer thicknesses, gradual increase in PMMA T_g and decrease in PC T_g was observed. Merging off the two T_gs was evident as the layer thicknesses reached 10 nm or less. Figure 3.39 shows the changes in the T_gs as a function of PC and PMMA layer thickness. Conventionally, merging T_gs was noticed during miscibility of polymer constituents or chemical reaction of individual components during processing. Considering the two immiscible components in the multilayered films and a short film processing time of 3 min, it is unlikely to cause any chemical reaction or transesterification between PC and PMMA [56]. Additionally, the convergence of the T_gs was not symmetrical [55].

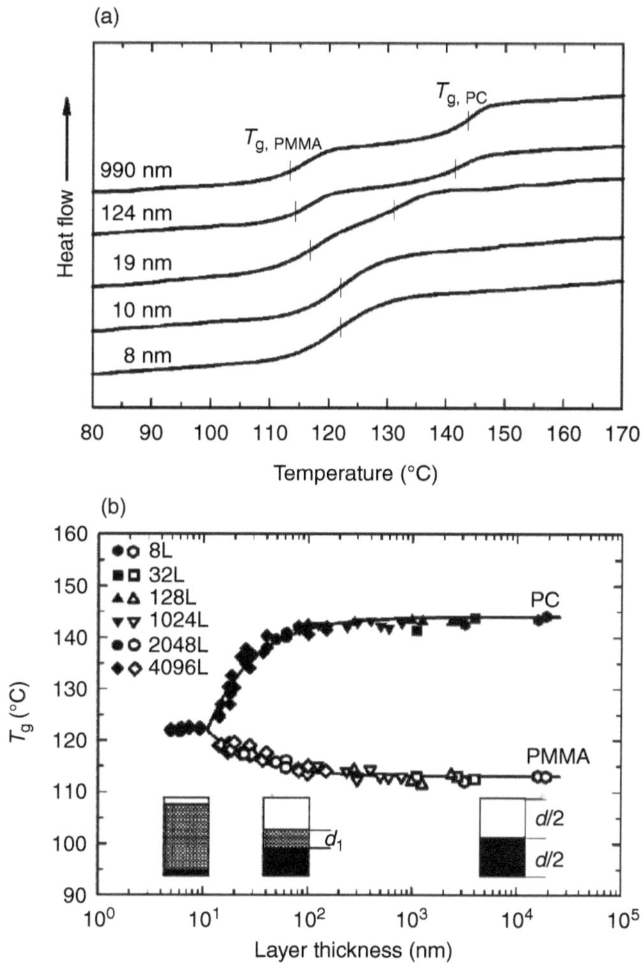

Figure 3.39 Glass transition behavior of PC/PMMA nanolayer films. (a) Heating thermograms show convergence of the heat capacity inflections corresponding to T_gs of the constituent polymers into a single inflection as the layers become thinner and (b) the dependence of T_g on layer thickness.

Using Fox law to calculate the T_g of the miscible blends, the estimated T_g of PC and PMMA blend was 126°C, 4–6°C higher than the measured T_g of 120–122°C of the nanolayered film. Increased T_g values were attributed to the increased density of nanolayered films due to changes in free volumes at nanolayer thicknesses. Further film characterization included measurements of various properties such as free volume size and oxygen permeability. The properties of PC/PMMA films with 10 nm layers were compared with the estimated values calculated using conventional laws and

Table 3.5 PP, HDPE, and Tie Layer Materials Used by Kamdar et al.

Materials	Density (g/cm^3)	M_w (kg/mol)	M_w/M_n
PP	0.900	340	4.6
HDPE	0.961	1210	9.6
EO855	0.855	10	2.2
EO876	0.876	83	2.0
OBC	0.880	120	2.8

From Table 1, Ref. [34].

relationships used for miscible blends as shown in Table 3.5. Beyond T_g changes, densification of the interphase occurs due to reduced free volume hole size rather than a decrease in the number of free volume holes. Effect of nanolayer thickness also showed interesting effect on the oxygen transport properties. The estimated permeability of PC/PMMA composites with layer thicknesses above 100 nm followed a series model for permeability for stacked layers, $1/P = \phi_{PC}/P_{PC} + \phi_{PMMA}/P_{PC}$, where ϕ is the volume fraction of PMMA or PC. However, the observed permeability of nanolayered films increased from 0.72 cm^3 (additive model) to 1.62 cm^3 (STP) cm/m^2/day/atm as seen in Figure 3.40. The permeability changes are significantly less than 1.92 cm^3 (STP) cm/m^2/day/atm estimated from the miscible blend model, $\ln P = \phi_{PC} \ln P_{PC} + \phi_{PMMA} \ln P_{PC}$. It was shown that the permeability changes occurred in the diffusivity, while the solubility remained unchanged.

This formation of a new phase (interphase) was demonstrated with properties different than the individual components, immiscible blends or miscible blend estimates. Based on the results, the interphase thickness was measured as 10 nm, in the nanolayered composite. The interphase thickness for PC/PMMA system was also calculated from Helfand and coworkers' theoretical predication. For immiscible polymers with infinite molecular weight, the correlation between interphase thickness and interaction parameter is defined as:

$$d_I = \frac{2b}{(6\chi)^{0.5}} \tag{3.5}$$

where b is the statistical segment step length and χ is the interaction parameter, which is estimated using solubility parameters and is given as,

$$\chi = \frac{V}{RT}\left(\delta_{PC} - \delta_{PMMA}\right)^2 \tag{3.6}$$

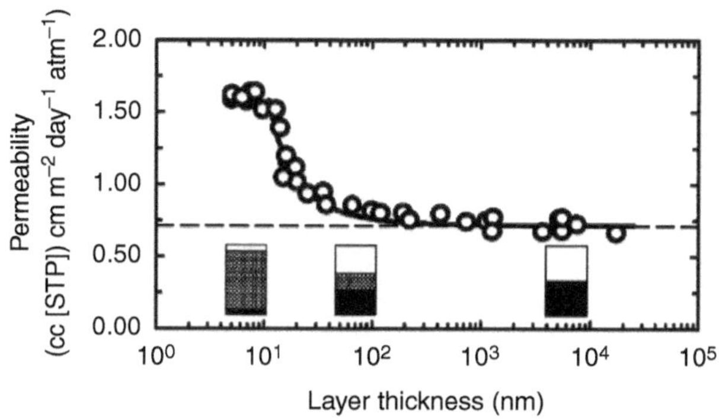

Figure 3.40 The effect of layer thickness on oxygen permeability of PC/PMMA films.

where V is the molar volume, T is the processing temperature, and δ_{PC} and δ_{PMMA} are 19.6 and 19.9 $(J/cm^3)^{1/2}$ [57]. The calculated interphase thickness for PC/PMMA layers at processing temperature of 543K was 9 nm.

It is evident that as the PC and PMMA layer thicknesses decreased, the interphase fraction becomes significantly higher, eventually leading to entirely interphase composite at 10 nm or less layer thicknesses. The overall properties shown in Table 3.5 for 10 nm layered films are the interphase material properties for PC/PMMA system. Further complex film assemblies were fabricated by systematically varying the relative amounts of PC and PMMA [58]. When PMMA layers were made as thin as 12 nm and the PC layer thickness was maintained at few tens of nanometers, the formation of interphase utilized entire fraction of PMMA leading to the formation of PC/interphase films. Based on film analysis, it is obvious that the PC/PMMA multilayered films are a three component system- PC, PMMA, and interphase layers. The schematic representation of the three-phase model is shown in Figure 3.41 [58].

Using oxygen permeability as a probe for multilayered films, interphase thicknesses for various other polymer pair systems were also calculated. The relationship between interphase thickness and interaction parameter showed that the interphase thickness was directly proportional to $\chi^{-0.5}$ as shown in Equation (3.5). A comparison of effect of interaction parameter on the interphase formation in two film systems, PMMA/PS (nominal layer thickness = 60 nm) and PC/PETG (nominal layer thickness = 45 nm) is shown in Figure 3.42 [59]. PMMA/PS film with an interphase thickness of 6 nm showed sharp layer boundaries, whereas PC/PETG with interphase thickness of 27 nm showed diffuse layers. These two systems and many other polymer combinations showed changes in

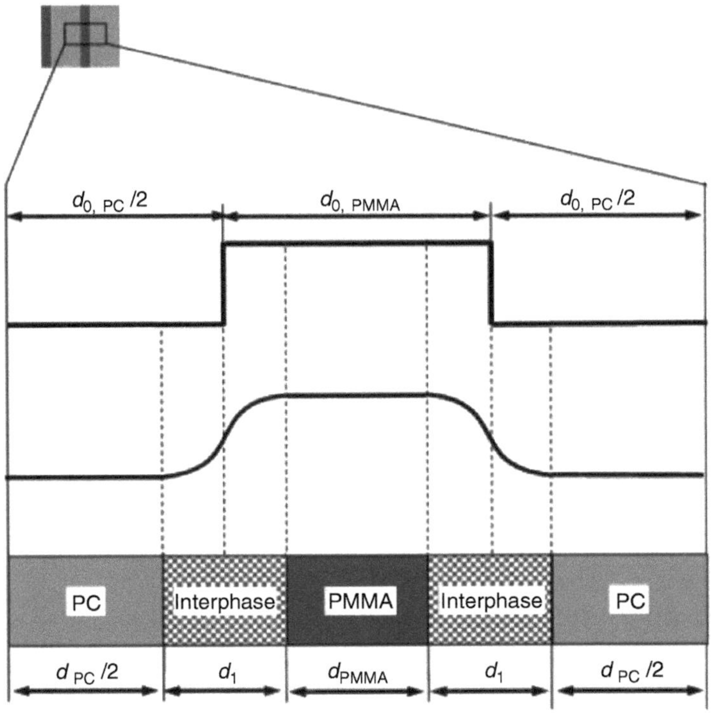

Figure 3.41 Schematic representation of the three-layered interphase model.

the oxygen permeability with increasing interphase fraction. Among other properties, increased interphase thickness also improved the interfacial strength. As the interphase thickness increased, number of chain entanglements increased resulting in an improved interfacial strength [59].

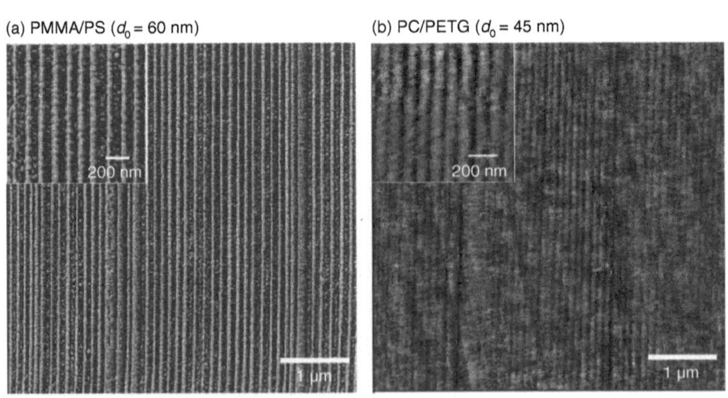

Figure 3.42 AFM phase images of film cross-sections . (a) PMMA/PS with nominal layer thickness of 60 nm; and (b) PC/PETG with nominal layer thickness of 45 nm.

The relationship between interphase thickness and the interfacial strength (G_{IC}) was established as $G_{IC} \propto d_I^2$ for polymer pairs with low G_{IC} [60]. Properties of various multilayered composites with variable interaction parameters were investigated to determine interphase properties as shown in Table 3.6.

Physical properties of multilayered films can be predicated based on the individual layer thicknesses and the interphase thickness. General observations in the multilayered films with interphases are:

- For individual layer thicknesses (d_0) > $10d_I$, films showed composition dependent additive properties of the two components.
- For $d_I < d_0 < 10d_I$, significant interphase fraction due to noticeable interfacial mixing of polymer chains showed impact on physical properties.
- For $d_0 < d_I$, completely interphase materials formed (Table 3.7).

3.5 Interdiffusion in Multilayered Polymers

Multilayer coextrusion process uses laminar flow conditions of combination of two or more polymers with short processing times producing composites and films with large number of interfaces between the coextruded polymers. In immiscible polymer systems or polymers with partial miscibility, composite structures with improved delamination and toughness were created by increasing number of interfaces. The ability to create thousands of interfaces also allowed processing of miscible polymers and controlling the interdiffusion. As the mixing of the two miscible polymers is significantly minimized during coextrusion process, due to short contact time between the composite polymers, controlled diffusion of the two polymers by subsequent thermal treatment allowed exploring the layer interdiffusion process. Effect of the number of multiplying dies on the interdiffusion of the two miscible polymers is schematically shown in Figure 3.43. When two polymers are brought together, movement of the polymer chains across the interface results into interdiffusion. The rate of disappearance of the concentration gradient across the interface, mutual diffusion coefficient D, is dependent on the composition of the system and it ranges between 10^{-16} m^2/s and 10^{-10} m^2/s for miscible polymer systems [61]. Coextrusion process was used to investigate the diffusion coefficient in multilayered films. Effect of interdiffusion on the thermal as well as gas transport properties was also investigated.

Table 3.6 PC/PMMA Interphase Properties and Estimated Properties From Various Models[a]

Properties	Units	PC	PMMA	PC/PMMA (10 nm Layers), Measured	Estimations	Differences
Glass transition	°C	144	112	122	126[b]	−4
Density	g/cm³	1.1934	1.1923	1.1945	1.1928[c]	0.0017
Free volume size	Å	2.82 ± 0.01	2.71 ± 0.01	2.71 ± 0.01	2.76 ± 0.01[d]	0.05
Oxygen permeability	cm³ (STP) cm/m²/day/atm	9.91	0.371	1.62	1.92[e]	−0.3

[a]*Properties of PC and PMMA monolayered films reported in Table 3.6 were used to calculate the estimated values as explained here.*

[b]*Calculated using Fox law for miscible blends,* $1/T_g = w_{PC}/T_{gPC} + w_{PMMA}/T_{gPMMA}$, *where* w_{PC} *and* w_{PMMA} *are the weight fractions of PC and PMMA, in multilayered films.*

[c]*Calculated using additive model,* $\rho = \rho_{PC}\phi_{PC} + \rho_{PMMA}\phi_{PMMA}$, *where* ρ_{PC}, ρ_{PMMA}, *are the densities of PC and PMMA;* ϕ_{PC}, ϕ_{PMMA} *are the volume factions of PC and PMMA;* ρ *is the calculated density of the 50/50 PC/PMMA films.*

[d]*Calculated using additive model, similar to density calculations.*

[e]*Using miscible blend model,* $\ln P = \ln P_{PC}\cdot\phi_{PC} + \ln P_{PC}\cdot\phi_{PMMA}$, *P is the permeability of the blend and* P_{PC} *and* P_{PMMA} *are the permeability values of PC and PMMA.*

Table 3.7 **Properties of Discrete Layers and Interphase Materials**

Constituent Polymers	Processing Temperatures (°C)	$\chi(10^{-3})$	d_I (nm)	P_d	P_h	P_I	ρ_d (g/cm³)	ρ_I (g/cm³)	r_d (Å)	r_d (Å)	G_{IC} (J/m²)
PMMA/PS	260	45	6.1	0.710	2.32		1.1187	1.1161	2.80	2.85[a]	45 ± 11
PMMA/SAN25	260	0.28	33	0.679	1.12	1.26	1.1316	1.1299	2.72	2.74	
PMMA/SAN30	260	2.4	8.0	0.639	0.988	1.12	1.1344	1.1324	2.72	2.76	80 ± 10
PC/PMMA	270	2.3	10	0.720	1.92	1.62	1.1928	1.1945	2.76	2.71	
PC/PETG	260	0.45	27	1.43	2.73	2.71	1.2365	1.2389	2.67	2.63	
P/SAN25	260	4.5	7.5	5.01	5.77	6.30	1.1346	1.1402	2.79	2.70	83 ± 15
PETG/PS	240	47	3.8	1.43	3.30		1.1624	1.1677	2.71	2.63	15 ± 5
PETG/SAN17	240	2.9	8.9	1.32	2.03	1.91	1.1732	1.1790	2.64	2.55	61 ± 10
PETG/SAN25	240	2.4	11	1.23	1.59	1.56	1.1796	1.1805	2.63	2.59	70 ± 10
PETG/SAN30	240	7.3	6.9	1.17	1.40	1.38	1.1815	1.1902	2.63	2.50	41 ± 6

χ, interaction parameter; d_I, interphase thickness; P_d, permeability of discrete layers in cm³ (STP) cm/m²/day/atm; P_h, permeability of homogeneous mixture in cm³ (STP) cm/m²/day/atm given by, where P_A and P_B are the permeabilities of the two polymers A and B; P_I, permeability of interphase in cm³ (STP) cm/m²/day/atm; ρ_d, density of discrete layers; ρ_I, density of interphase; r_d, free volume hole radius of discrete layers; r_I, free volume hole radius of interphase; G_{IC}, interphase strength.

[a]*Film with 84% interphase.*

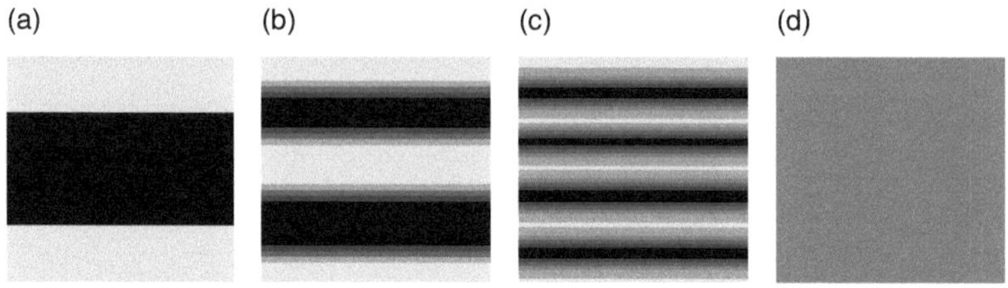

Figure 3.43 Coextrusion process depicting the interdiffusion process with increasing number of layers during layer multiplication process. The layers multiplied, as shown in the images (a–d). Schematic (a) represents sharper layer boundaries, while (b) and (c) represent increased interdiffusion. The interlayer diffusion process created concentration gradient and the interdiffusion increased with increasing layers eventually creating homogeneous blend as shown in (d). This process can also be visualized in thermal annealing of coextruded films with miscible polymers.

3.5.1 Polycarbonate and Copolyester Composites

Multilayered composites of PC/KODAR (a copolyester of 1,4-cyclohexanedimethanol and terephthalic acid) were produced with 657, 1313, 1857, and 3713 layers and 60/40 (wt/wt) composition. Additional samples of 657- and 1313-layered composites with 40/60 (wt/wt) composition were also fabricated in this study [62]. The individual layer thicknesses ranged from 0.6 μm to 3.7 μm. Two distinct T_gs of PC and KODAR were observed in the layered composites. Annealing of the composite structure at 200°C led to intermediate glass transitions between the T_gs of the two components, suggesting some interdiffusion. With further increase in annealing time, merging of the two glass transitions into a single T_g after 4 h showed complete interdiffusion of PC and KODAR layers in 657-layered composite (PC thickness, 3.7 μm; KODAR thickness, 2.6 μm) as shown in Figure 3.44.

The ability of the polymers to interdiffuse under controlled thermal treatment, as well as the symmetry of the multilayered composite allowed composition profile modeling using Fick's equation. One dimension, nonsteady state diffusion concentration profile was defined by Fick's equation as:

$$\frac{\partial W_i}{\partial t} = D\frac{\partial^2 W_i}{\partial x^2} \tag{3.7}$$

where, W_i is the weight fraction of species i, t is the diffusion time, x is the position, and D is the mutual diffusion coefficient. The interdiffusion element required to solve Equation (3.7) is defined as one-half of PC layer, one-half of adjacent KODAR layer and the interface. Only on-half layers

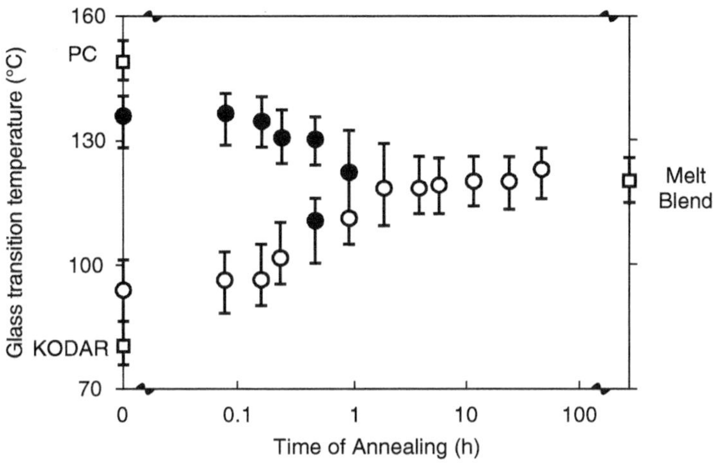

Figure 3.44 The T_gs and breadths of the PC/KODAR 60/40 composite with 657 layers as a function of the annealing time at 200°C.

of the two components were considered due to symmetry of the layered composite. Many assumptions equivalent to the case of weakly interacting polymer pairs were used in this analysis. The mutual interdiffusion coefficient was assumed to be a function of temperature and independent of the composition. Additionally, the position of the interface between the diffusion species was assumed to be constant. Assuming initial sharp interface and composition gradient at the boundaries of interdiffusion element as zero, Equation (3.7) can be solved by method of separation of variables resulting into,

$$W_{PC}(x,t) = \frac{L_{PC}}{L_{PC}+L_K} + \sum_{n=1}^{\infty} \frac{2}{n\pi} \sin\left(\frac{\pi n L_{PC}}{L_{PC}+L_K}\right) \cos\left(\frac{2\pi n x}{L_{PC}+L_K}\right) \exp\left(\frac{-4\pi^2 n^2 D t}{\left(L_{PC}+L_K\right)^2}\right) \quad (3.8)$$

where W_{PC} is the weight fraction of PC, L_{PC} and L_K are the layer thicknesses of PC and KODAR.

Arbitrary values of D and t were used to achieve the composition profile. The composition profile obtained from Equation (3.8) for $m = 15$ resulted into accuracy of 0.001%. All other parameters in Equation (3.8) are known from the composite structure. The composition profile was then converted to T_g using Fox Equation (3.9):

$$\frac{1}{T_{g,PC}(x,t)} = \frac{W_{PC}(x,t)}{T^0_{g,PC}} + \frac{1 - W_{PC}(x,t)}{T^0_{g,K}} \quad (3.9)$$

Detailed analysis by changing D values was performed to obtain the best fit to glass transition data. It was observed that the T_gs merged faster with increasing D values. It is intuitive that higher mutual diffusion coefficient will increase the miscibility of the two components at faster rates. The model also accounted for the diffusion occurred during the coextrusion process itself. In temperature range from 200°C to 230°C, the values of D varied from 4.0×10^{-16} m²/s to 1.6×10^{-16} m²/s, which were close to reported mutual diffusion values of 10^{-12}–10^{-15} m²/s for other miscible polymer pairs [63]. It is interesting to note that when the annealing temperature was decreased to 195°C, the rate of interdiffusion was comparable with the crystallization rate of KODAR. This complex behavior was further studied by Nazarenko et al. [64].

3.5.2 Nylon-EVOH Composites

Coextrusion approach was further used to investigate interdiffusion behavior of nylon and EVOH multilayered composites [65]. The effect of the ethylene content in EVOH was investigated on the interdiffusion properties with nylon. Nylon was coextruded with two different grades of EVOHs with different amounts of ethylene – EVOH44 (44 mol% ethylene) and EVOH24 (24 mol% ethylene). Interdiffusion of nylon and EVOH was revealed by morphological analysis shown in Figure 3.45. Well-defined layer boundaries with sharp interfaces were observed in 17-layered films, and 129-layered films showed diffused interfaces, but the two polymer components were still visible in this sample. As the number of layers increased to 1025, completely interdiffused nylon/EVOH layers were observed. The rate of the interdiffusion in nylon/EVOH44 film was slower than the nylon/EVOH24 system as evident from the sharper boundaries in 129-layered nylon/EVOH24 film. It is evident that in the conventional nylon/EVOH composite structure a very few number of layers will show distinct nylon and EVOH layers.

Oxygen permeability was used to calculate the diffusion coefficient in this system unlike T_g in PC/KODAR system. By extending Equation (3.8) to this system, the composition profile in nylon/EVOH composites was converted to permeability profile according to this equation for miscible blends.

$$\ln P(x,t) = \varphi_N(x,t)\ln P_N + \left(1-\varphi_N(x,t)\right)\ln P_E \qquad (3.10)$$

where ϕ_N is the volume fraction of nylon, P_N and P_E are the permeability values of nylon and EVOH, respectively. The permeability of the interdiffusion element was calculated at different values of D to establish the best

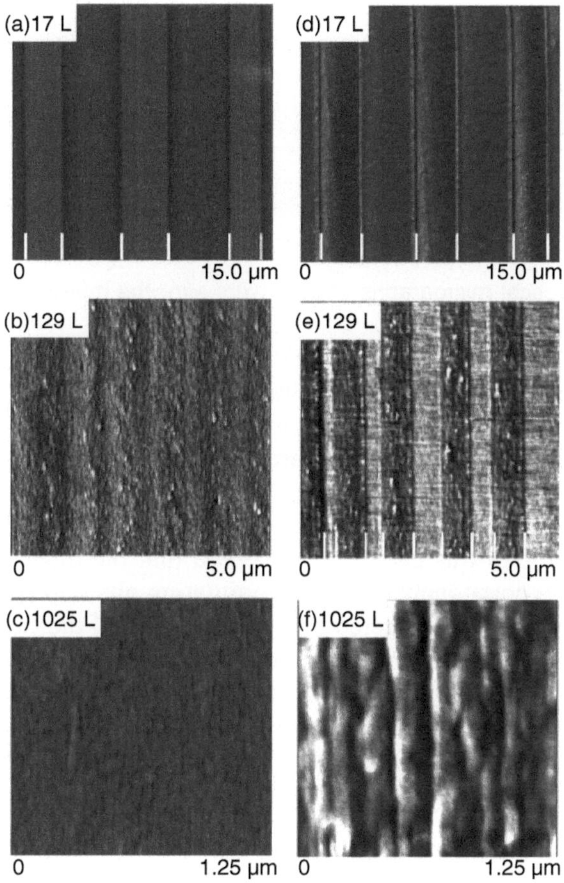

Figure 3.45 AFM phase images showing the layer morphology in films with varying number of layers for (a–c) nylon / EVOH44 multilayers and (d–f) nylon/EVOH24 multilayers. (a,d) 17 layers with nominal layer thickness of 3.2 μm; (b,e) 129 layers with nominal layer thickness of 400 nm, and (c,f) 1025 layers with nominal layer thickness of 50 nm.

correlation between model and experimental values. The best D values for nylon/EVOH multilayered films were 2×10^{-12} m²/s for EVOH with 44 mol% ethylene and 2.4×10^{-13} m^2/s for EVOH with 24 mol% ethylene content. Combination of modeling and experimental data was used to establish stronger interaction of nylon with EVOH44 than with EVOH24.

3.5.3 Interdiffusion in Polyethylenes

The approach of multilayer coextrusion was further extended to explore interdiffusion of linear and branched polyethylene [66]. Diffusion of highly

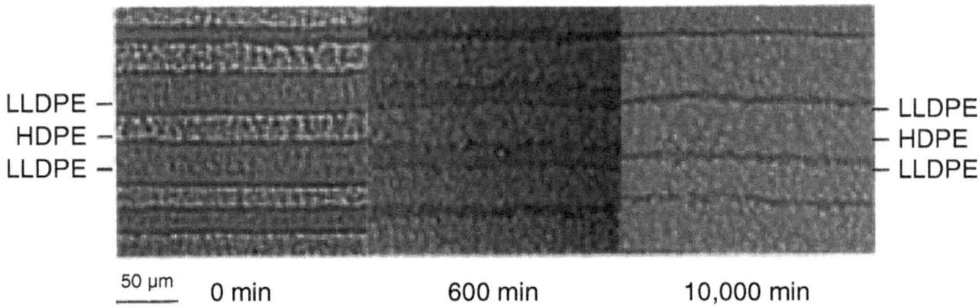

Figure 3.46 Optical micrographs of the layers after the multilayers were taken to 200°C for the time indicated.

mobile chains into a region of less mobile chains, results in osmotic pressure, which causes the layer boundaries to move in the direction of more slowly diffusing chains [67]. An example of HDPE/LLDPE film annealing at 200°C and its effect on moving boundary is shown in Figure 3.46. The shrinkage in the LLDPE layer indicated the moving boundary only at high annealing times. In this case, it is observed that the diffusion kinetics of mobile LLDPE components was the dominating phenomena. The interdiffusion was significantly affected by polydispersity of the two components. The interlayer boundaries did not change significantly at short interdiffusion times, but only moved when high molecular weight fractions were involved in the interdiffusion process. Interdiffusion in miscible polymers was further exploited to concentrate the inorganic particles in one of the components in multilayer composites as discussed in Chapter 5.

3.6 Summary

Multilayered films and composites with hundreds of layers produced confined nanostructures of commercial polymers in continuous film production process. Improvements in the gas barrier properties by aligning the crystalline lamellae at few tens of nanometer thicknesses allowed improvements in the barrier properties of PEO and PCL. Controlled crystallization of polymer nanolayers under the confining substrate in the layered composite was also used to change the lamellar orientation for gas barrier improvement as demonstrated in sPP and PVDF polymers. Use of multilayered structures in biaxially oriented PP and PET demonstrated capabilities to extend the use of confined structures as gas barrier films.

Multilayering also showed its impact on adhesion and mechanical properties in various polymer composites. Adhesion, deformation, fatigue, and failure properties showed shift in the corresponding mechanisms with increasing number of layers and changing layer thicknesses. Extending the approach to filled, gradient and BCPs showed ability to manipulate the composite properties at nanoscale, which are otherwise difficult to control in conventional blends. The layer thicknesses and compositions can be tuned to achieve the desired mechanical properties.

Two other key phenomena, interphases, and interdiffusion, play important roles in defining the gas barrier and mechanical properties. Furthermore, these phenomena are important in defining the polymer blend properties in many commercial applications. Multilayer coextrusion and subsequent thermal treatment successfully explored interphases and interdiffusion in different polymer pairs. Interphase properties, previously difficult to characterize, were studied successfully using conventional analysis techniques. Entirely interphase materials using immiscible polymers were produced in multilayered films with hundreds to few thousand layers. Controlled thermal annealing of the multilayered film allowed observing interdiffusion in polymers and estimating diffusion parameters.

References

[1] W.J. Koros, Barrier Polymers and Structures, ACS symposium series 423, ACS, Washington DC, (1990) (Chapter 1).
[2] H.P. Wang, J.K. Keum, A. Hiltner, E. Baer, B. Freeman, A. Rozanski, A. Galeski, Science 323 (2009) 757–761.
[3] H.P. Wang, J.K. Keum, A. Hiltner, E. Baer, Macromolecules 42 (2009) 7055–7076.
[4] P.H. Hermans, Contribution to the Physics of Cellulose Fibres, Elsevier, Amsterdam, 1946, pp. 195.
[5] E.L. Cussler, S.E. Hughes, W.J. Ward III, R. Aris, J. Membr. Sci 38 (1988) 161.
[6] H.P. Wang, J.K. Keum, A. Hiltner, E. Baer, Macromolecules 43 (2010) 3359–3364.
[7] (a) H. Schonherr, C.W. Frank, Macromolecules 36 (2003) 1188. (b) S. Vyazovkin, J. Stone, N. Sbirrazzuli, J. Therm. Anal. Calorim. 80 (2005) 177.
[8] H.P. Wang, J.K. Keum, A. Hiltner, E. Baer, Macromol. Rapid Comm. 31 (2010) 356–361.
[9] M. Ponting, Y. Lin, J.K. Keum, A. Hiltner, E. Baer, Macromolecules 43 (2010) 8619–8627.
[10] J.M. Carr, D.S. Langhe, M.T. Ponting, A. Hiltner, E. Baer, J. Mat. Res. 27 (10) (2012) 1326–1350.
[11] (a) A.M. Chatterjee, F.P. Price, J. Polym. Sci. Polym. Phys 12 (1975) 2368. (b) A.M. Chatterjee, F.P. Price, J. Polym. Sci. Polym. Phys. 13 (1975) 2385. (c) A.M. Chatterjee, F.P. Price, J. Polym. Sci. Polym. Phys. 13 (1975) 2391.
[12] D.S. Langhe, A. Hiltner, E. Baer, Polymer 52 (25) (2011) 5879–5889.
[13] M. Mackey, L. Flandin, A. Hiltner, E. Baer, J. Polym. Sci. B 49 (2011) 1750–1761.

[14] S.J. Pan, J. Im, M.J. Hill, A. Keller, A. Hiltner, E. Baer, J. Polym. Sci. B 28 (7) (1990) 1105–1119.
[15] G. Zhang, P.C. Lee, S. Jenkins, J. Dooley, E. Baer, Polymer 55 (2014) 663–672.
[16] (a) S. Tomoyuki, T. Takaharu, Polymer 34 (2002) 891–896. (b) K. Hiroshi, T. Minor, C. Yozo, T. Hiroyuki, J. Polym. Sci. 16 (1978) 131–142.
[17] (a) T. Lupke, S. Dunger, J. Sanze, H.J. Randusch, Polymer 45 (2004) 6861–6872. (b) Y.J. Lin, P. Dias, H.Y. Chen, A. Hiltner, E. Baer, Polymer 49 (2008) 2578–2586.
[18] (a) M.R. Kamal, H. Garmabi, S. Hozhabr, L. Arghysis, Polym. Eng. Sci. 35 (1995) 41–51. (b) J.T. Yeh, W.H. Yao, Q.C. Du, C.C. Chen, J. Polym. Sci. Part B 43 (2005) 511–521.
[19] Y. Lin, A. Hiltner, E. Baer, Polymer 51 (2010) 4218–4224.
[20] Y. Lin, A. Hiltner, E. Baer, Polymer 51 (2010) 5807–5814.
[21] (a) P. Chandran, S. Jabarin, Adv Polym. Technol. 12 (2) (1993) 153–165. (b) R.Y.F. Liu, Y.S. Hu, D.A. Schiraldi, A. Hiltner, E. Baer, J. Appl. Polym. Sci. 94 (2004) 671–677.
[22] J.M. Carr, M. Mackey, L. Flandin, A. Hiltner, E. Baer, Polymer 54 (2013) 1679–1690.
[23] (a) M.A. Osman, V. Mittal, H.R. Lusti, Macromol. Rapid. Comm. 25 (2004) 1145–1149. (b) T.C. Bissot, in: W.J. Koros (Ed.), Barrier Polymers and Structures, American Chemical Society, Washington DC, 1990, pp. 225.
[24] M. Gupta, Y. Lin, T. Deans, E. Baer, A. Hiltner, D.A. Schiraldi, Macromolecules 43 (2010) 4230–4239.
[25] D. Jarus, A. Hiltner, E. Baer, Polymer 43 (2002) 2401–2408.
[26] D. Jarus, A. Hiltner, E. Baer, Polym. Engg. Sci. 41 (2001) 2162–2171.
[27] (a) W. Heckmann, G. E. McKee, F. Ramsteiner, Mechanical Properties of Polymers Based on Nanostructure and Morphology, in: G.H. Michler, F.J. Balta-Calleja (Eds.), 2005, pp. 435, (Chapter 11). (b) C. Harrats, G. Groeninckx, Mechanical Properties of Polymers Based on Nanostructure and Morphology, in: G.H. Michler, F.J. Balta-Calleja (Eds.), 2005, pp. 487, (Chapter 11).
[28] T. Ebeling, A. Hiltner, E. Baer, J. Appl. Polym. Sci. 68 (1998) 793–805.
[29] T. Ebeling, A. Hiltner, E. Baer, Polymer 40 (1999) 1985–1992.
[30] J. Kerns, A. Hsieh, A. Hiltner, E. Baer, J. Appl. Polym. Sci. 77 (2000) 1545–1557.
[31] M. Nishimoto, H. Keskula, D.R. Paul, Polymer 32 (1991) 272–278.
[32] V. Ronesi, Y.W. Cheung, A. Hiltner, E. Baer, J. Appl. Polym. Sci. 89 (2003) 153–162.
[33] (a) D.J. Arriola, E.M. Carnahan, P.D. Hustad, R.L. Kuhlman, T.T. Wenzel, Science 312 (2006) 714–719. (b) L.P.C. Sham, L.G. Hazlitt, Macromol. Symp. 257 (2007) 80–93. (c) F.C. Stehling, T. Huff, C.S. Speed, G. Wissler, J. Appl. Polym. Sci 26 (1981) 2693–2711. (d) D.W. Barlett, J.W. Barlow, D.R. Paul, J. Appl. Polym. Sci. 27 (1982) 2351–2360.
[34] A.R. Kamdar, R.K. Ayyar, B.C. Poon, G.R. Merchand, A. Hiltner, E. Baer, Polymer 50 (2009) 3319–3328.

[35] P. Dias, Y.J. Lin, B. Poon, H.Y. Chen, A. Hiltner, E. Baer, Polymer 49 (2008) 2937–2946.
[36] T. Ebeling, S. Norek, A. Hasan, A. Hiltner, E. Baer, J. Appl. Polym. Sci. 71 (1999) 1461–1467.
[37] B.C. Poon, S.P. Chum, A. Hiltner, E. Baer, Polymer 45 (2004) 893–903.
[38] T.R. Bernal-Lara, A. Ranade, A. Hiltner, E. Baer, Mechanical Properties of Polymers Based on Nanostructure and Morphology, in: G. H. Micher, F.J. Balta-Calleja (Eds.), pp. 629–681, 2005.
[39] J. Im, W.J. Schrenk, J. Plastic Film Sheet. 4 (1988) 104–115.
[40] J. Im, E. Baer, A. Hiltner, High Performance Polymers, in: E. Baer, A. Moet (Eds.), Hanser Publishers, pp. 175–198.
[41] M. Ma, K. Vijayan, A. Hiltner, E. Baer, J. Mat. Sci. 25 (1990) 2039–2046.
[42] D. Haderski, K. Sung, J. Im, A. Hiltner, E. Baer, J. Appl. Polym. Sci. 52 (1994) 121–133.
[43] J. Kerns, A. Hseih, A. Hiltner, E. Baer, A. Appl. Polym. Sci. 77 (2000) 1545–1557.
[44] R. Adhikari, V. Seydewitz, K. Loschner, G.H. Michler, A. Hiltner, E. Baer, Macromol. Symp. 290 (2010) 156–165.
[45] R. Adhikari, W. Lebek, R. Godehadrt, S. Henning, G.H. Michler, E. Baer, A. Hiltner, Polym. Adv. Technol. 16 (2005) 95–101.
[46] US Patent No. 4,540,623, 1985.
[47] M. Ma, J. Im, A. Hiltner, E. Baer, J. Appl. Polym. Sci. 40 (1990) 669–684.
[48] C.D. Mueller, S. Nazarenko, T. Ebeling, T.L. Schuman, A. Hiltner, E. Baer, Polym. Eng. Sci. 37 (2) (1997) 355–362.
[49] M. Ponting, T.M. Burt, L.T.J. Korley, J. Andrews, A. Hiltner, E. Baer, Ind. Eng. Chem. Res. 49 (2010) 12111–12118.
[50] C. Lai, R. Ayyar, A. Hiltner, E. Baer, Polymer 51 (2010) 1820–1829.
[51] C.-Y. Lai, A. Hiltner, E. Baer, L.T. Korley, ACS Appl. Mater. Interf. 4 (4) (2012) 2218–2227.
[52] T.M. Burt, J. Keum, A. Hiltner, E. Baer, L.T.J. Korley, ACS Appl. Mater. Interf. 3 (2011) 4804–4811.
[53] T.M. Burt, A.M. Jordan, T.J. Korley, ACS Appl. Mater. Interf. 4 (2012) 5155–5161.
[54] (a) E. Helfand, A.M. Sapse, J. Chem. Phys. 62 (1975) 1327–1331. (b) E. Helfand, Y. Tagami, J. Polym. Sci. Polym. Lett. 9 (1971) 741–746. (c) E. Helfand, Y. Tagami, J. Chem. Phys. 56 (1972) 3592–3601.
[55] R.Y.F. Liu, Y. Jin, A. Hiltner, E. Baer, Macromol. Rapid Comm. 24 (16) (2003) 943–948.
[56] M. Rabeony, D.T. Hseih, R.T. Garner, D.G. Peiffer, J. Chem. Phys. 97 (1992) 4505.
[57] D.W. Van Krevelen, Properties of Polymers, Elsevier, Amsterdam, 1990, 189 (Chapter 7).
[58] R.Y.F. Liu, A.P. Ranade, H.P. Wang, T.E. Bernal-Lara, A. Hiltner, E. Baer, Macromolecules 38 (2005) 10721–10727.

[59] R.Y.F. Liu, T.E. Bernal-Lara, A. Hiltner, E. Baer, Macromolecules 38 (2005) 4810–4827.
[60] R.Y.F. Liu, T.E. Bernal-Lara, A. Hiltner, E. Baer, Macromolecules 37 (2004) 6972–6979.
[61] M. Tirrell, H.H. Kausch, Annu. Rev. Mat. Sci. 19 (1989) 341–377.
[62] G. Pollock, S. Nazarenko, A. Hiltner, E. Baer, J. Appl. Polym. Sci. 52 (1994) 163–176.
[63] (a) C. Macahte, J. Lohmar, M. Droscher, Macromol. Chem. 191 (1990) 3011. (b) R.S. Porter, J.M. Jonza, M. Kimura, C.R. Desper, E.R. George, Polym. Eng. Sci. 29 (1989) 55.
[64] S. Nazarenko, D. Hardeski, A. Hiltner, E. Baer, Macromol. Chem. Phys. 196 (1995) 2563–2585.
[65] Khariwala D, PhD 2009. Thesis, Structure-property relationships in multi-layered polymeric system and olefinic block copolymers.
[66] T. Schuman, E.V. Stepanov, S. Nazarenko, G. Capaccio, A. Hiltner, E. Baer, Macromolecules 31 (1998) 4551–4561.
[67] P.F. Green, C.J. Palmstrom, J.W. Mayer, E. Kramer, Macromolecules 18 (1985) 501.

4 Optical Properties of Multilayered Films

4.1 Introduction

Traditional discussions of polymer, layered or homogenous, material optical properties would be centered on the transparency, haze, or color of the polymer materials. For nearly 75 years, chemists and engineers have attempted to control polymer material optical properties through chemical structure alterations, incorporation of chemical additives along the lines of clarifier molecules, blending with high optical property secondary materials, or processing the material quickly to quench crystallization. This is a notorious hindrance to material transparency and/or minimizes the material thickness in final products. Multilayer processing of films, sheets, or profiles comprised of hundreds or thousands of layers, will also benefit from these scientific foundational advances in controlling material or final product optical properties; however, their highly ordered, submicron-layer thickness adds an additional tool into the engineer's kit for tailoring or, in unique cases, inducing new product optical properties.

This chapter will demonstrate cases where the optical properties of multilayered articles with layer thicknesses less than 1 μm, usually <500 nm, have been produced in the scientific and commercial marketplace over the last 50 years. The nanolayered materials exhibit novel optical effects, which are not achievable in homogenous or blend polymer systems, due to a synergistic combination of contrasting material properties and confinement in highly ordered films and sheets comprised of hundreds or thousands of repeating layers. Examples of novel optical property layered films systems and technologies include optically reflective nanolayered films, nonlinear optically absorbing films, nanolayered lasing films, optically patternable films, and variable refractive index materials. Additionally, technological applications enabled by some of these nanolayered film optical properties will be discussed including gradient optical lenses, flat prisms, waveguides for solar cells, and laser beam shaping devices.

4.2 Polymeric Reflector (Bragg) Films

Realization by the DOW chemical company in the early 1960s that it could produce polymer films with tens or hundreds of layers via its system of sequential layer multiplying dies, described in Chapter 1, enabled first

Manufacturing and Novel Applications of Multilayer Polymer Films
http://dx.doi.org/10.1016/B978-0-323-37125-4.00004-6

experimentally produced polymeric films with structural induced optical reflections or coloring. These early nanolayered film, produced by Alfrey et al. [1], comprised of 201 repeating layers of two different polymer materials with a significantly different refractive index, that is difference in material refractive index (Δn) was 0.1 or greater. Alfrey et al. translated relationships previously utilized in inorganic vapor deposited periodic systems to produce a mathematical predictive model in order to describe the case of a nanolayered thin polymer film, where two colorless polymer materials with a differing refractive index can be coextruded into a nanolayer structure. It appeared to be colored, tinted, or optically reflective due to internal film structural reflections with nanolayers of similar size scale to the wavelength of UV, visible, or IR light waves.

Directly related to a light reflection phenomena rooted at the nanolayer film interface of both the materials with a difference in refractive index, visible reflections from multilayered film structures are produced when the individual polymer layer thicknesses approaches 100 nm. This photonic phenomenon has been previously modeled and reported for cases of uniform [1] and graded [2,3] thickness layered polymeric films. This mathematical model first defines the incremental phase angle, Φ_i, contributions of each individual polymer layer as,

$$\Phi_i = \frac{2\pi n_i d_i}{\lambda}$$

where d_i and n_i represent the layer thickness and refractive index, respectively. A matrix approach describing the optical behavior of each layer can be expressed as,

$$\left({}_i q\right) = \begin{pmatrix} {}_i q_{11} & {}_i q_{12} \\ {}_i q_{21} & {}_i q_{22} \end{pmatrix} = \begin{pmatrix} \cos\Phi_i & j\dfrac{\sin\Phi_i}{n_i} \\ jn_i \sin\Phi i & \cos\Phi_i \end{pmatrix}$$

where $j = \sqrt{-1}$. Through definition of each layer thickness, d_i, and refractive index, n_i, any two component multilayer film layer thickness distribution can be described by a complex 2×2 matrix (Q), which represents the matrix product of N individual layer matrices (NQ):

$$(Q) = \left({}_1 q\right)\left({}_2 q\right)\left({}_3 q\right)\left({}_4 q\right)\cdots\left({}_N q\right)$$

The multilayered film reflection spectrum for normal incident light in an air, that is refractive index medium of 1.0, is then expressed as

$$R = \left[\frac{(Q_{11} + Q_{12}) - (Q_{21} - Q_{22})}{(Q_{11} + Q_{12}) + (Q_{21} + Q_{22})} \right]^2$$

Utilizing this model as a guide, the effect of varying a nanolayered film layer thickness distribution and/or constituent polymer material refractive index can be investigated as a means to develop or predict the optical reflective properties based solely on structural reflections of light passing through the layered object. Verification of the model predictive capabilities is published in even recent literature [4] with an example of a 128-layered polystyrene/poly(methyl methacrylate) (PS/PMMA) film simulated and experimentally measured optical reflection spectra shown in Figure 4.1. The 12 μm thick layered film comprised of 128 individual layers each 90 nm thick and processed utilizing sequential layer multiplying dies after coextrusion from a two layer A/B feedblock. As seen in Figure 4.1b, visible light spectrum transmission minima, or a peak reflection, was simulated to lie between 530 nm and 560 nm, which was in good agreement with the experimentally measured 540 nm minima peak with 5% transmission and 95% reflectivity. It should be noted that differences between the simulated and measured film transmission spectra, outside of the alignment of the minima peak, were witnessed at 500 and 600 nm. This deviation resulted from an inhomogeneous nanolayer thickness distribution taken from an AFM film cross-section. From the AFM image, a single standard deviation of the measured layer thickness distribution was in the order of 10–15% for 90 nm target. Variations or deviations on the scale of 5% or greater will result into the appearance of these secondary or side peaks in the reflective film spectrum.

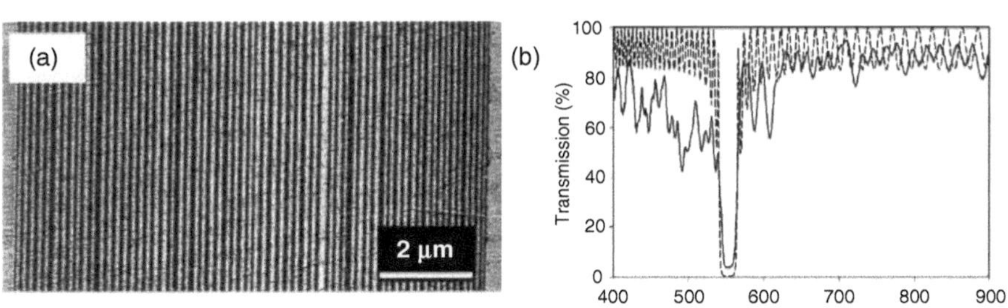

Figure 4.1 (a) A 128-layered PS/PMMA multilayered film AFM cross-section showing a uniform 90 nm layer thickness distribution. (b) Simulated (dashed) and experimentally measured (solid) reflection spectra showing a 40 nm wide, high intensity reflection peak centered on 560 nm.

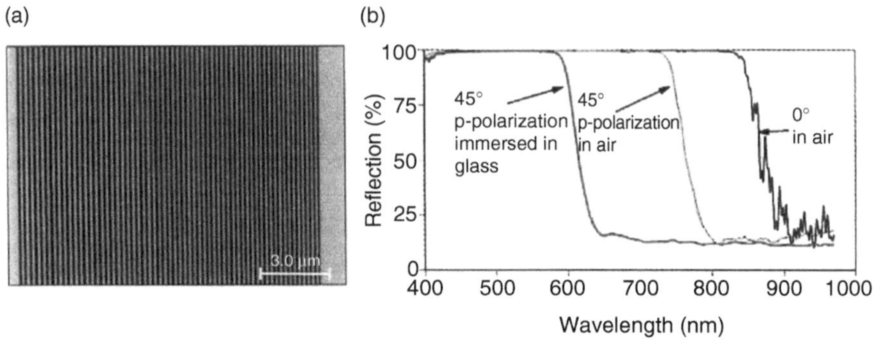

Figure 4.2 (a) AFM image of a birefringent optics stack gradient layers of PMMA and polyester nanolayers. The film possesses layers approximately 25% thicker across the image from left to right. (b) Optical reflection of a 448-layered PMMA/polyester film. Film employs a designed gradient layer thickness to expand the reflection bandwidth to 500 nm.

Holding a minimal deviation in the final layered substrate thickness distribution is paramount for achieving the highest levels of layered structure optical reflection. An alternative processing approach to the sequential layer multiplying die process is to directly coextrude a film from a multilayered feedblock, as illustrated in Figure 4.2, which shows a multilayered film with alternating layers of PMMA and polyethylene terephthalate (PET) with a layer thickness gradient of 25%. Modern tool and die making equipment is capable of producing tool control to submillimeter dimensions that enable production of feedblocks with up to a thousand layers. The fabrication of these hundred- to thousand-layered feedblocks enables a direct comparison of the optical effect of higher levels of flow induced layer nonuniformity in sequential layer multiplier approach, as compared to the single direct feedblock coextrusion approach. To illustrate the optical effects of this layer thickness deviation, a case study looking at the visible reflective spectrum of two gradient layer thickness multilayered films follows.

A highly reflective nanolayered film exhibiting a 500 nm wide reflection band was produced through direct feedblock coextrusion of a 448 gradient layer film [5]. This film exhibited a sharp transmission drop to 15% at 900 nm in air (Figure 4.2b). In contrast, a similar 128-layered gradient layer reflective film was coextruded utilizing a series of six nonuniform layer multiplying dies [6] to produce a film structure with a 1.2 μm–110 nm layer thickness distribution, as shown in Figure 4.3. This 128-layered gradient layer film, again exhibits a 12% single sigma layer thickness distribution that results in higher than simulated transmission, or lower levels of

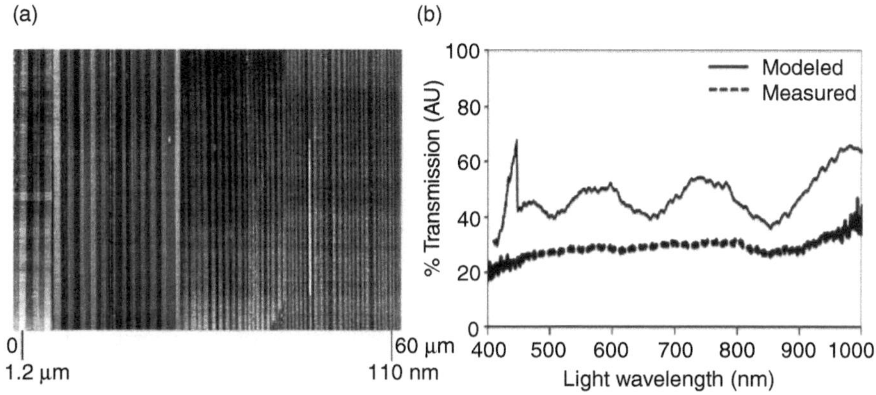

Figure 4.3 (a) AFM image and (b) reflectivity spectrum of a 10× gradient 128-layered PS/THV multilayered film with 1–100 nm layer thicknesses.

reflectivity, and the appearance of periodicity or local minima secondary reflection peaks. In comparing the two films, it is important to acknowledge that both polymer processing techniques were successful at producing gradient layer thickness distribution films with wide band, that is, >500 nm wide reflections. The larger layer thickness distribution of the layer multiplication processing technique clearly produces inferior optical reflection properties to the direct feedblock approach. The realization that a clear superiority of optical film performance in direct nanolayer coextrusion feedblock processing techniques begs a question as to what motivation exists to perpetuate the use of combined feedblock and sequential layer multiplying die coextrusion approaches to producing optical layered films.

One answer to this questions lies in balancing performance, capital costs, and product flexibility in a layered film coextrusion line. In the previous example of gradient layered optical films, a reflective film was produced through coextrusion directly from a feedblock machined with 100s of layers and exhibited a strong transmission dropoff to 15% for 900 nm and longer wavelength light. This result better matched the optical reflection spectrum models as compared to the secondary peaks witnessed in the layer multiplying die-produced film. An advantage of the layer multiplication die approach lies in its ability to adapt to the process of producing a different number, or in the case of gradient layer thickness films, layer thickness distributions, without the need to acquire additional capital tools. Figure 4.4 illustrates another 128-layer gradient layer thickness film fabricated from a series of six nonuniform layer multiplying dies. Details of gradient layer multipliers and their film production process have been discussed in Chapter 2.

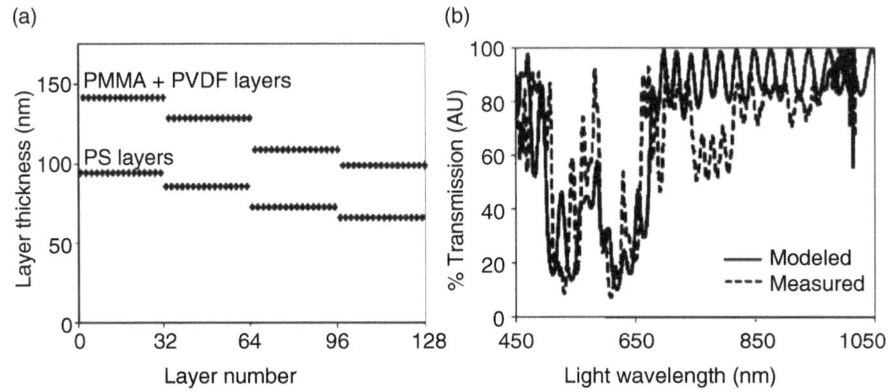

Figure 4.4 (a) Unconventional dual step gradient layer thickness distribution through the thickness of a 128 gradient layer PS/PMMA + PVDF blend layer film. (b) Dual reflection peak spectra, simulated and experimentally measured, with a high band-pass at 580 nm for the dual peak 128 gradient layer multilayered film.

In contrast to Figure 4.3, the shape of an internal layered film thickness distribution got drastically changed by changing the multiplier die sequence order, yet that the same layer multiplying dies were utilized for coextrusion processing trials in Figures 4.3 and 4.4. The dual-tiered gradient layer thickness film distribution [6] illustrated and coextruded into the optically reflective gradient-layered film, resulted in a drastically different film reflection spectra. The optical reflection resembles a dual peak reflection filter with an intermediate high transmission band pass zone. This drastic change in optical film reflection spectra is easily achieved through reorientation of the layer-multiplying die; however, this would require the machining of an entirely different layered feedblock to achieve. The economics and downtime required to change outer production line feedblocks could be prohibitive to produce two different gradient layer film distributions, while developing markets or applications might not support this level of investment. It is in these new product development areas where layer multiplying dies hold an advantage over direct micro- or nanolayered feedblock coextrusion methods. Both techniques are valid engineering approaches capable of producing submicron multilayered optical films, as demonstrated in application examples as discussed in this chapter, both techniques have been or are currently being utilized to fabricate nanoscale films and products, with truly unique optical properties.

4.3 Optically Active Nanolayered Film Systems

The first applications leveraging the unique optical reflection of coextruded micro- and nanolayered, polymeric Bragg crystal films were nontechnical, consumer decorative applications. As previously discussed, DOW discovered the optical reflection effects of films with pluralities of submicron films in the 1970s. Although never producing commercial products with the technology, DOW licensed the production of optically reflective films to the Mearl Corporation, in the 1970s, and eventually the 3M Corporation, 1990s, who both began leading producers and innovators in leveraging the technology into consumer goods and unique technological applications. The Mearl Corporation, later acquired by Engelhard and BASF, first commercialized micro- and nanolayered reflective films for consumer decorative applications. Products included Mearl's prismatic polyester gift bag paper [7,8]. Cryovac product laminate sheets and shrink wrap sleeves for bottles [9], optically detectable tamper resistant bottle seals [10], and Mearl Corporation dye doped layered structures exhibiting a pearlescent appearance for gift packaging bows [11]. 3M Corporation product innovations in the 1990s developed the wide optical reflection brand gradient layer thickness films for optical filtering [12], Vikuiti branded reflective and surface textured light repolarization and brightness enhancement inserts for LCD display monitors, IR, and/or UV reflective films, as architectural energy saving window wrap films [13].

Successful commercialization of the microlayered optical film technology by companies such as Mearl and 3M Corporations, have led to significant advanced research efforts to investigate additional consumer and technologically advanced uses for the optically reflective microlayered polymer films. Innovations to nearly 30 years old microlayered optically reflective film field, can be traced back to 1989 at the DOW Chemical Company, with exhibited structurally reflecting films in reversibly deformable elastomeric films [14]. These "stretching" microlayered films would be, for example, coextruded to a designated color appearance, red or green; however, they would undergo a visibly apparent color shift to blue wavelengths with an externally initiated mechanically stretching or elongation. Upon release of this tensile stress, assuming the applied strain was below the nonrecoverable plastic material deformation threshold, the blue appearance of these stretched films would revert back to the red or green color, as the film layers elastically recovered and thickened back to the extruded dimensions. An example of the strain-based optical shifting elastomeric reflective film [4] is demonstrated in Figure 4.5, where the optical reflection band, or reflective film color/appearance, can be demonstrated

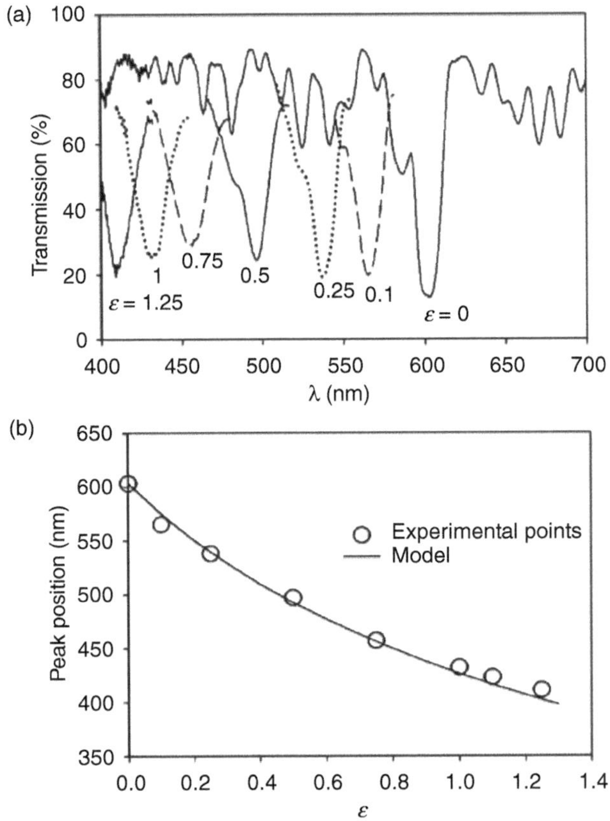

Figure 4.5 (a) Effect of strain on 128-layered polyurethane/copolyester elastomeric multilayered film reflection properties. (b) Relationship of 128-layered polyurethane/copolyester film peak reflection plotted as a function of overall film deformation strain.

as a smooth function of applied tensile deformation strain. This type of reversible reflective film has been identified as a potential passive optical based sensor for strain or pressure-/bladder-type products.

Another example of recent advances in initially commercialized nanolayered reflective films involves the incorporation of active optical dyes or molecules into layered film structures to enable additional optical functionality beyond reflection or light filtration. Though introduction of colored chemical dye molecules into the nanolayered reflective films date back to the Mearl Corporation [8] in 1987, these first films solely utilized the dye molecules to accent reflective film color appearance, hue, or intensity. A new class of dye doped nanolayered optically reflective films appeared in the technical literature in the late 2000s that leveraged the nanolayered film structure as an optical scaffold, to impregnate and/or selectively disperse functional molecules, dyes, or inorganic particles, which when exposed to an external optical stimulator, produced a novel optical response at the film level. Further sections review and describe these optically active nanolayered film responses that include: nonlinear reflective properties, optical lasing, or three-dimensional (3D) optical patternability.

A proposed method toward the demonstration and development of an optical intensity dependent optical filter or shield was reported through the introduction of a nonlinear optical dye material into a periodically repeating nanolayer, in a two and three component coextruded optical film [15]. Armstrong et al. described the nonlinear absorbance properties obtained by creating a 128-layered cast film comprising of alternating layers of polyester terephthalate glycol and polycarbonate loaded to 4 wt% with a nonlinearly absorbing optical dye, lead tetracumylphenoxy phthalocyanine, as shown in Figure 4.6. Inclusion of the nonlinear optical dye in polycarbonate layers was intended to introduce an intensity-dependent optical filter to an already optically reflecting Bragg-like film.

In addition to the previously described Bragg-crystal structural reflection, induced from intrafilm layer thicknesses and individual layer refractive index differences, the nanolayered film could exhibit a secondary reflective behavior, if exposed to high intensities of dye-exciting light wavelengths. Activation of the optical dye via a high intensity monochromatic light source, such as in the case of exposure to an optical laser, would excite and added intralayer reflective mechanism resulting in filtration of the high intensity wavelength when the exposure is above the molecular dye activation levels. Armstrong et al. demonstrated the ability to not only disperse and maintain nonlimiting dye functionality in the nanolayered film, but also restrict entropic driven dye migration from PC to PETG through introduction of a mass transport fluoropolymer barrier layer. Restricting the physical migration of active molecules is a key to maintain optical functionality, as the dye is defunctionalized in the PETG polymer matrices. The ability to restrict optical additive movement across micro- or nanolayer boundaries to maintain optical dye functionality and efficiency is a key in this and the following examples of nanolayered optical lasing films.

Nanolayered reflective films were coextruded with lasing dye molecule doped layers to demonstrate two distinct architectures of an all polymer laser film device (Figure 4.7). An all-plastic distributed Bragg reflector (DBR) laser was demonstrated using coextruded 128-layered PS/PMMA reflective films, laminated onto a laser dye filled polymer film [16].

The nanolayered DBR laser operates on an absorption, wavelength conversion, and emission operation model that is dependent on the specific

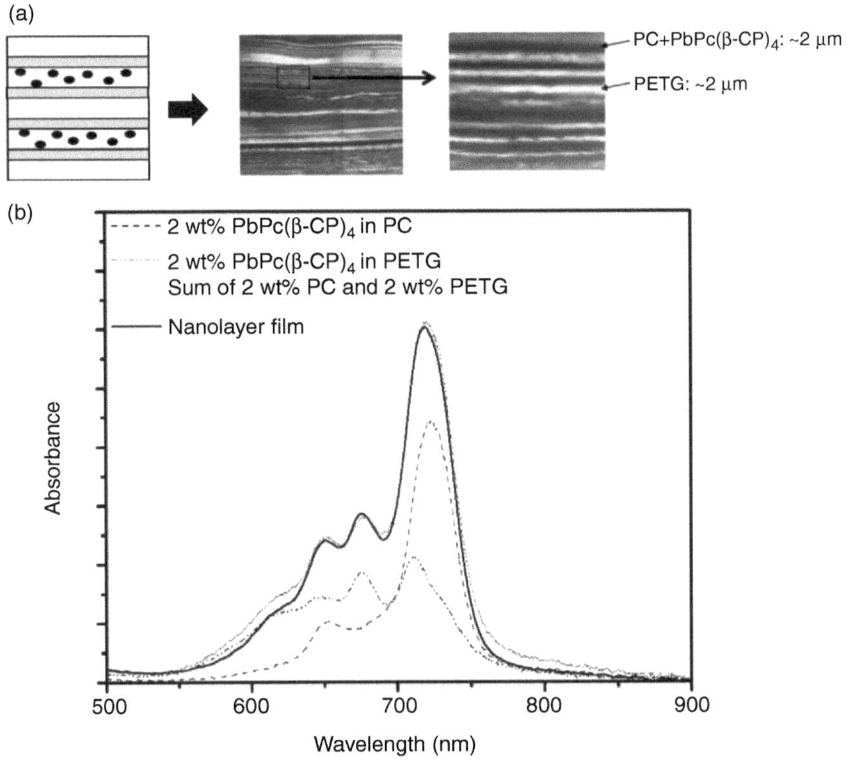

Figure 4.6 (a) Schematic illustrating 129-layered NLO dye loaded polymer film system. Film thickness = 11 μm and 86 nm NLO dye filled layers. The micrograph of coextruded nanolayered PC + NLO dye layers with alternating PETG layers without NLO dye has been shown. (b) Optical absorbance of coextruded nanolayered film with NLO as compared to conventional dye loaded material.

absorption and excitation spectra of the lasing dye contained with the its dye filled gain layer. To activate the lasing film, the dye doped gain layer is optically pumped using a nanosecond Nd:YAG laser at a wavelength matching the dye absorption spectrum. The surrounding multilayered reflectors are designed such that the absorption wavelength of the lasing dye is outside the reflector film main reflection band. Additionally, the peak fluorescence of the lasing dye layer is designed to fall within the peak reflector film reflection band, to magnify and align the lasing signal. The optical pump is therefore transparent through the DBR structure, to enable activation and excitation of the optical lasing dye, thereby inducing an optical emission through the polymer reflector layers. The DBR laser was demonstrated with a variety of lasing dyes; rhodamine 6G (R6G),

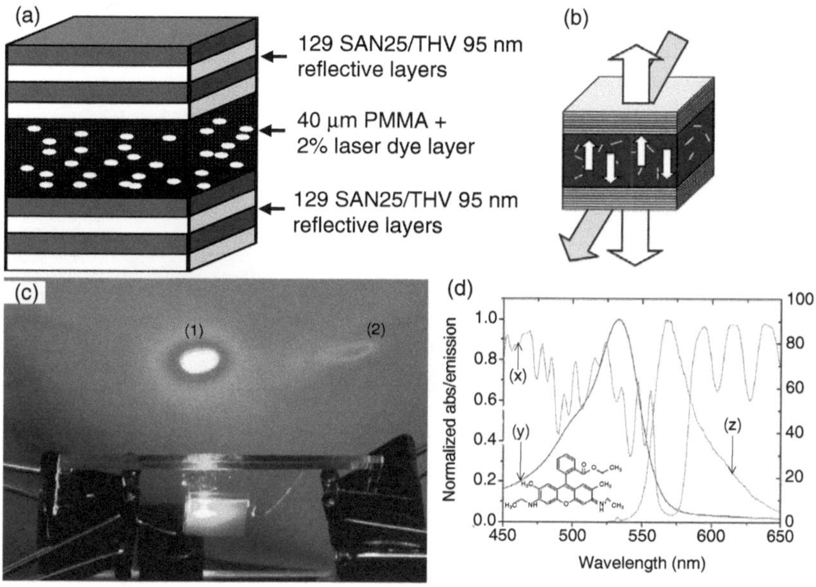

Figure 4.7 (a) Schematic of DBR nanolayered mirrors sandwiched between a laser dye-doped gain media. (b) Optical lasing of SAN25/THV nanolayered DBR laser with rhodamine laser dye doped PMMA gain media. (c) Bright emission from R6G DBR laser. The saturated bright spot on the screen (marked as (1)) is the lasing emission and the fainter green spot (marked as (2)) to the right is the residue of the pump light. (d) Reflection band (marked as (x)), absorption (marked as (y)) and emission (marked as (z)) of rhodamine (R6G).

pyyromethene 580 (P580), and 1,4-bis-(α-cyano-4-methoxystyryl)-2,5-dimethoxy-benzene (C1RG) with a maximum emission efficiency of 50% in the case of P580 dye. The ability to replace heavy, hard to process inorganic materials as the reflectors for a laser, could lead to much cheaper manufacturing path, especially when the entire all plastic laser device can be produce in a single coextrusion step, as demonstrated by the nanolayered distributed feedback (DFB) laser.

A second example of a lasing, nanolayered polymeric film device was produced utilizing a single coextrusion step to produce the gain and reflective functionalities in a single thin film. The DFB laser differs from the previously described DBR example in that an optical lasing dye is compounded directly into one of the periodic polymer film nanolayers, and the DFB structure comprises of both gain and feedback media [17]. The DFB laser film structure demonstrated in Figure 4.8 was constructed in a 12 μm thick film comprised of 128 layers [18]. The A/B

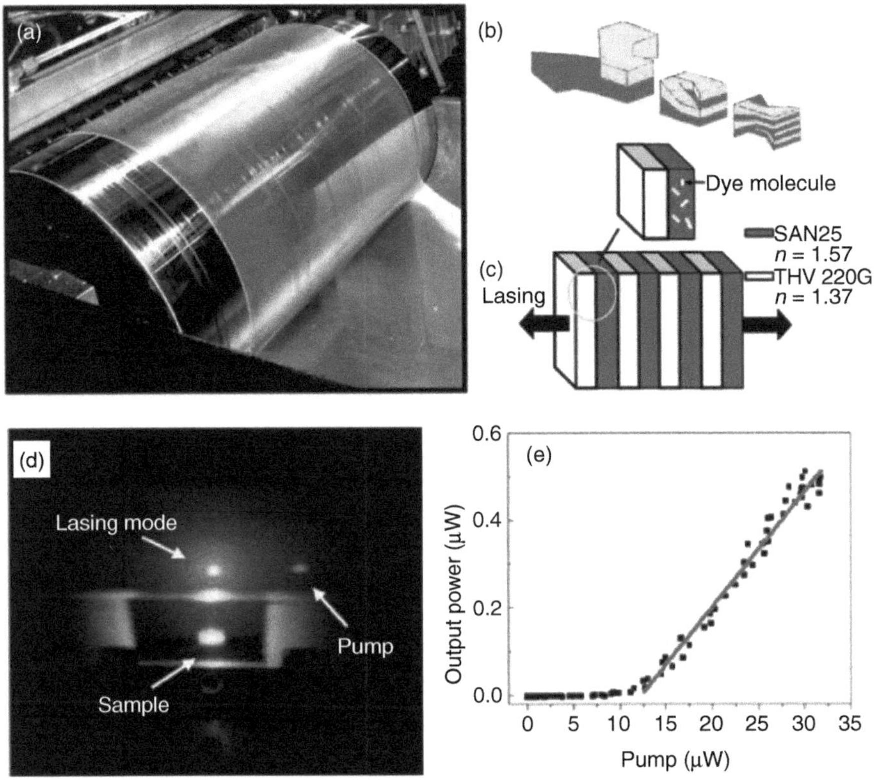

Figure 4.8 Schematic illustration of nanolayered laser dye loaded polymer film with 129 layers and 95 nm thick alternating SAN25, n = 1.57, and THV, n = 1.37, reflective layers. (a) Roll-to-roll processing of a C1-RG DFB laser film. (b) Details of the layer multiplication. (c) Details of the layered architecture. (d) Image of the laser action. (e) Pump versus average output power plot of a laser pumped at 50°; the slop efficiency is 2.6% and the lasing threshold is 238 μJ/cm^2.

repeating film layers comprised of (a) poly(styrene-*co*-acrylonitrile) containing 25 wt% acrylonitrile (SAN25, n = 1.56) doped with about 1 wt% R6G laser dye and (b) a fluoro-terpolymer consisting of tetrafluoroethylene (TFE) and hexafluoropropylene (HFP) (n = 1.37). The fluoro-terpolymer was selected for this application based on a twofold need for (a) a large Δn needed to induce optical reflection and (b) due to its low chemical solubility with the selected lasing dyes. The chemical barrier, similar to the case of the nonlimiting optical films previously described, ensured that the active optical dye molecules did not migrate across the layer boundaries and reduce device efficiency as a result of entropic diffusion.

Operation of the DFB laser, similar to the case of the DBR, requires dye activation via absorption of an external laser, emitted at a wavelength to excite emission from the optical dye, which corresponds with the optical reflection band of the nanolayered film. The advantage of the DBR laser is its ability to mass process a low cost device in a single coextrusion step, resulting in mass producible rolls of laser material requiring no additional fabrication before excitation and use, however, at the cost of a reduced efficiency, 2.6%, as compared to the Bragg layers. The ability to produce working laser film devices is directly paramount in controlling and eliminating the dye migration across the nanolayer film boundaries, an ability which can be leveraged by other optical film uses where exact 3D location of optical dye molecules can lead to volumetrically writable optical films.

4.4 Photopatternable Films

Photoreactive polymers have been utilized as masking or templating materials by the inorganics and semiconductor industry for decades. The ability to photoresist or pattern functionalized polymer materials represents a pathway to realization of security enhanced films, optical data storage media, and nanotemplated material reactors or "lab on a chip"–type devices. Recent development efforts to incorporate photopatternable materials into coextruded micro- and nanolayer films has demonstrated a mass producible, economically optimized route to produce these type of devices, as demonstrated through recently published examples. Photopatternable films were demonstrated via a photoactivated reaction, to initiate a change in a single layer refractive index in a 1024-layered coextruded film [19] cast to 100 μm thickness, as shown in Figure 4.9. The A/B structured nanolayered film comprised of 100 nm thick individual layers of (a) homogenous polymethylmethacrylate (PMMA) and (b) a blend of PMMA with 12 wt% trans-cinnamic acid (CA). The refractive index of the as-blended PMMA/CA layer was measured at 1.511. Exposure of the blend material to 254 nm UV light resulted in a dimerization of CA that results in a drop in the material refractive index to 1.504. As-extruded layered films of the PMMA, $n = 1.490$, and CA blend layers, $n = 1.511$, possess a significant refractive index difference ($\Delta n = 0.02$) to appear visually cloudy or hazy. Upon exposure to UV light, the refractive index of CA blend layer is shifted from $n = 1.511$ to $n = 1.504$, as a result of a photoreactive dimerization in the CA. The reduction in CA blend layer RI is sufficient to lessen the refractive index difference at layered film interfaces ($\Delta n = 0.007$), enabling UV exposed areas of layered film to appear transparent. Through

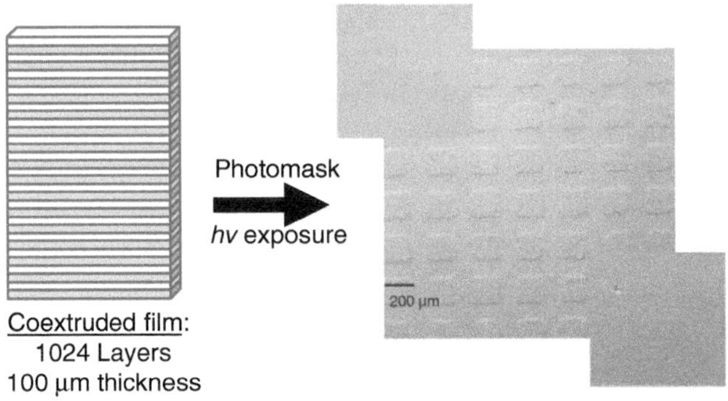

Figure 4.9 Schematic illustrating coextrusion approach to produce a nanolayered photopatternable PMMA with 12wt% CA/ PMMA cast film. Micrographs display actual measured optically induced photopatterned 100-μm film after UV exposure. (Upper left) PMMA/PMMA control film, (center) PMMA with 12% CA layered with PMMA, and (lower right) PMMA with 12% CA blend.

the mechanism of dimerization in CA blend layers, conducted by UV light, a template nanolayer film was demonstrated by Tangirala et al. [19].

As with the previous examples in this text, the reader may ask, "why nanolayers" in thinking about the intricate material and processing design required to process films as compared to a standard three layer film. At first glance the Tangirala example should be achievable in layered films with only three layers, the mechanism for photopatterning is to match the refractive index through the bulk of the film. A top and bottom layer of PMMA with a CA blend layer will exhibit the same result as the nanolayered film in achieving a reduced Δn at the layer interface. Additionally, even at 1024 layers, a structural reflection from the additive contributions of the interfaces would still be small with a $\Delta n = 0.02$ to affect the film transparency noticeable, so why were all the layers necessary? A follow-up example utilizing a selective depth approach to "write" unique information into each subsequent layer through the film thickness can provide the answer.

Three-dimensional photopatterning of nanolayered films was demonstrated by Ryan et al., through the coextrusion of a 64-layered film, with individual writable layers loaded with 2 wt% of a fluorescent chromophore 1,4-bis(α-cyano-4-octadecyloxystyryl)-2,5-dimethoxybenzene dye (C18-RG) [20]. The coextruded film was of an A/B structure comprised of individual (a) layers 300 nm thick of PETG loaded with 2 wt% C18-RG and (b) 3.1 μm layers of a fluoropolymer. The (a) layers represented 32 patternable layers, while the (b) layers acted as a diffusion barrier for the C18-RG dye as well as spacing material to allow for resolution between

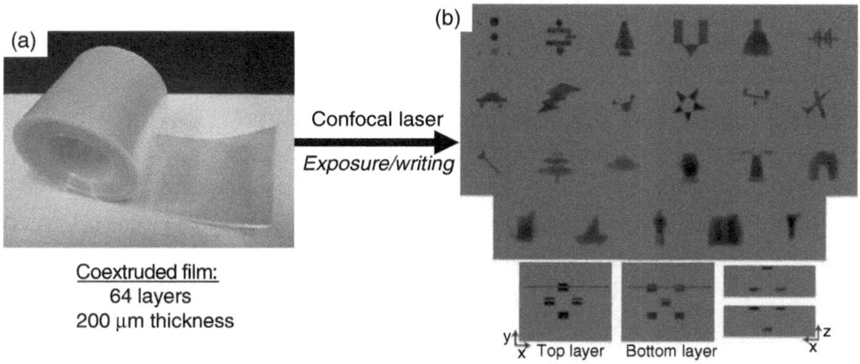

Figure 4.10 Schematic illustrating coextrusion approach to produce 3D writable film. (a) Multilayered film roll containing 64 layers (b) image micrograph displays 23 unique patterns etched into 23 consecutive PETG + dye nanolayers through the coextruded film thickness. Each image is 22 × 22 μm in size. Cross-section images of the patterned film also shown at the bottom of part of (b).

the written layers. Once coextruded, the overall 110 μm thick-layered film was photopatterned to record unique information in 23 or the 32 available PETG + dye layers through the film thickness via scanning confocal microscopy utilizing a 405 nm CW laser. An image of the unique data patterned into the 23 layers through the film thickness is shown in Figure 4.10. The photopatterning process was stable, with features reported resolvable after 2 years from the date of experiment.

The ability to record unique data through nanometer spaced domains in the multilayered film could enable terabyte level storage capabilities in the patternable film based devices in the near future. The enabling technology of multilayer coextrusion process to position patternable dye filled layers, repeatedly at nanometer precise intervals, through the film thickness highlights a unique opportunity for optical photopatterning films with tens, hundreds, or even thousands of layers. In addition to expanded data storage capacity with layers, the discretization of the nanolayers and separation with the fluoropolymer barrier layers allows maintained resolution during data writing due to the significant refractive index difference between layers, as well as performance, and material cost optimization by providing the dye diffusivity barrier due to the low chemical solubility. Utilizing the internal bulk of a nanolayered film for enhanced data storage, highlights a usage were the full volume of a polymer film is utilized to initialize an optical effect. The further section of this chapter highlights certain advantages of bulk versus surface material utilization, in the demonstration of refractive optical applications of nanolayered polymer films.

4.5 Nanolayer-Enabled Polymer Refractive Optical Devices

Reducing the individual layer thickness of polymeric films from millimeter into the micrometer and down to the 100 nm scale, results in a series of optically reflective Bragg crystal films and applications based on these diffractive technologies. Further section examines the optical effect of extending a layer scaling investigation of polymer material systems with a similar magnitude refractive index difference at the layer interface, $\Delta n \sim 0.08$. Case studies utilizing nanolayered optical film with very thin individual layer thicknesses (<50 nm) are discussed in the framework of refractive optical applications and demonstrators optics.

Polymer material systems, with sufficient differences in constituent material refractive index, arranged in alternating layers, possesses a refractive index modulation, whose period corresponds to an individual layer thickness [21]. When the wavelength of light is equal to (λ) or approaches one quarter ($\lambda/4$) of the film's individual layer thicknesses, the layered film material exhibits optical reflectivity, such as the structural Bragg grating reflections. However, when the intrafilm individual layer thickness approaches less than one quarter of the wavelength of light, $\lambda/4$ less than 75 nm, the periodicity of the layered polymeric materials becomes smaller/shorter to diffract the passing light and thereby results in the film exhibiting a pseudohomogenous bulk optical identity. For example, in a series 4096 layer, 50 μm thick layered PMMA and styrene-*co*-acrylonitrile, with 17% acrylonitrile content (SAN17) films [22], with a varied relative PMMA:SAN17 volumetric content ratio, resulting films exhibited a single, bulk film refractive index value (Figure 4.11).

Furthermore, this single refractive index point directly corresponded to the volumetric average of two constituent polymer materials, equation denoted in Figure 4.11. The physical appearance of these $< \lambda/4$ thickness layered films, as a result of the periodicity of the PMMA/SAN17-layered thickness being significantly shorter than the wavelength of visible light, is optically transparent with measured transmission values equal to that of constituent homopolymers. The optical transparency of nanolayered films remains at > 90% transmission in spite of a difference in constituent material refractive index of 0.08 for PMMA (1.489) and SAN17 (1.572) when measured at 633 nm. The series of 101 different volumetric combinations of PMMA/SAN17 films ranging in 1% volumetric increments from 0/100 to 100/0 represent a series of transparent optical films where the bulk film refractive index systematically varies by 0.0008. The production of optically transparent films with volumetrically dependent tunable

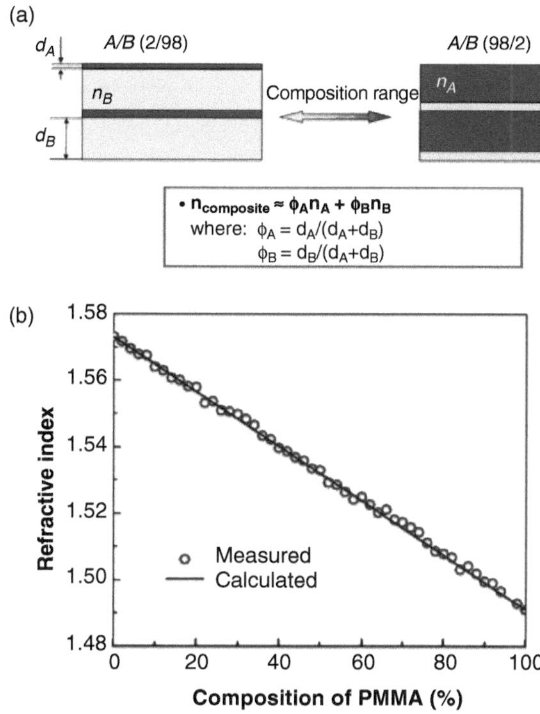

Figure 4.11 (a) Relationship of nanolayered film, with layer thickness $< \lambda/4$ of light, volumetric composition to composite film refractive index. (b) Experimentally coextruded and measured 4097-layered PMMA/SAN17 nanolayer film composite refractive index as a function of volumetric loading of PMMA.

film refractive index has potential usage as an index-matching film in solar cells, display technology, and has even been utilized as building block materials to produce custom gradient refractive index (GRIN) lenses.

GRIN materials are a class of material and refractive optic, which possess a distribution of material refractive index as one progresses from its surface through the bulk. The variation in internal material refractive index acts to continually bend light, as it is transmitted through the medium, a phenomena usually reserved for the curved interfacial surface of refractive optics, where the difference in air and polymer or glass lens refractive index provides the optical power. GRIN lenses are commonly found in nature, in animal, and human eye lenses [23], and serve to add optical power to lenses, while simultaneously correcting spherical and chromatic aberrations, and improving optical resolution of imaging systems in compact spaces. A cooperative research team out of Case Western Reserve

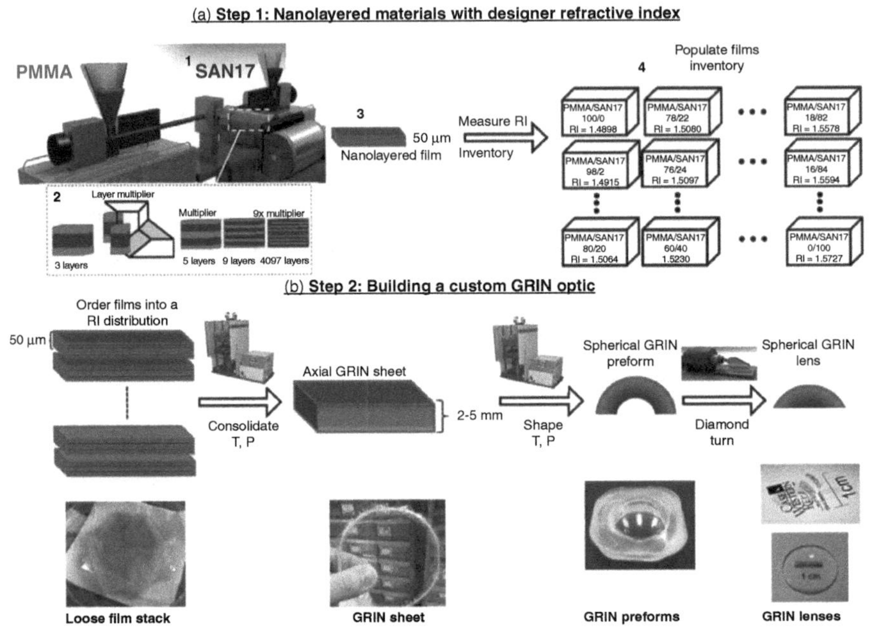

Figure 4.12 Nanolayer film enabled fabrication process to create custom 3D spherical GRIN optics.

University and Naval Research Laboratory have adapted series of nanolayered polymeric material systems, with individual layer thicknesses $< \lambda/4$, into a fabrication process to create high resolution refractive GRIN optical materials and lens systems [24–26].

The first step in the production of nanolayered polymeric GRIN lens fabrication, Figure 4.12, is the coextrusion of a series of 50-μm thick, 4096-layered PMMA and polystyrene-*co*-acrylonitrile with 17 mol% acrylonitrile (SAN17) films. The refractive index difference between PMMA (1.489) and SAN17 (1.572) is 0.083, which is proportional to the optical power of the GRIN system. Additional nanolayered material systems have been reported, which also utilize PMMA (1.489) and a fluoropolymer blend layer (1.43) or a polycarbonate (1.58) and an optical amorphous polyester material (1.63). Regardless of the polymer material system utilized, a series of films with a varied volumetric combination of polymer material *A* and *B* are varied at 1% steps, so their respective contributions were 100/0, 99/1, 98/2, and ended at 0/100. This series of 101 separate nanolayered films of PMMA/SAN17 are utilized as discrete building blocks, to construct GRIN distribution in a polymer media.

Constructing a refractive index gradient is accomplished by sequentially stacking individual 50-μm thick nanolayered PMMA/ SAN17 polymer films with differing bulk refractive indices. The nanolayered polymer film stacks can be arranged so that their refractive index profiles bear a linear shape, second-order polynomial shape, third-order polynomial shape, or any custom distribution requested or simulated in an optical lens design, as long as the refractive index distribution is bounded by a maximum and minimum index value, which is equal to the constituent polymer materials. Typically, stacks consist of between 75 and 300 individual nanolayered films. Once the desired refractive index distribution profile is stacked, typically in a clean room environment to prevent contamination, a thermoforming consolidation process is completed at elevated temperatures and pressures to lock-in the desired axial GRIN profile. This consolidated GRIN stack is usually a 3–50-mm thick flat sheet, which possesses an axial refractive index distribution normal to the surface plane. Subsequent shaping, polishing, or diamond turning steps in the consolidated axial GRIN sheet, enables the construction of axial, spherical, or radial GRIN optics. Converting an axial GRIN sheet to a radial or spherical GRIN distribution is accomplished by a second thermoforming operation. Within this operation, an axial GRIN sheet is formed into a meniscus-like lens, or GRIN preform. This is done by molding it into a concave–convex mold pair. Curved GRIN preforms can be further transformed into a variety of finished optics by additional diamond turning. Using this methodology, a variety of GRIN containing geometric lens shapes could be formed that include meniscus shaped, plano–convex, plano–concave, or plano optical lenses. Basically, any machinable, spherical, or aspheric shaped geometric optic can be fabricated with an internal GRIN distribution. Internal GRIN distribution shapes, or internal refractive index curvature, varies between axial, spherical, or special radial GRIN distribution. Due to the formability of polymer materials, GRIN optics can be fabricated with a large range of geometric radii and diameters. These radii could be as sharp as 3 mm and approach a nearly planar, infinite radius. GRIN optics with diameters ranging from 6 mm to 60 mm has been fabricated. As compared to conventional GRIN sol–gel deposition or interdiffusion techniques, the nanolayering approach provides fabrication freedom enabling breadth, not only in the shape of an optic's refractive index distribution (axial, radial, or aspheric), but also in the optical diameter, optic thickness, and surface curvature. Expansion of the traditional GRIN materials manufacturing limitation allows significant changes to the optical lens or imaging system design process through an additional degree of design freedom in the bulk material nanolayered GRIN distribution, to the correction of optical defect or poor performance in optical systems. GRIN optical lenses have been fabricated utilizing this polymer nanolayer film-enabled fabrication that includes: a bio-inspired polymeric copy of the human eye lens, a 3D Luneburg-like ball lens, and a significantly lighter 86% imaging optics for night vision goggle and 25% lighter rifle scope eyepiece optics. The nanolayered GRIN optics technology has the potential to miniaturize and light-weight existing zoom camera and imaging systems, which invokes an improved IOL implant performance, and produces some novel nonimaging optics that can change the way engineers envision optics in traditional applications, as discussed further.

Nanolayered films, based GRIN optics, were successful in improving imaging and camera lens system performances through the introduction of optical power into the bulk lens material. Stemming from the ability to discreetly construct custom shaped GRIN distributions and nonimaging GRIN applications have been demonstrated to showcase the novel optical functionality of nanolayered optical polymer materials. Prisms are traditionally cut as pyramids or facetted objects, with a varying optical path length, which is dependent on the entrance and exit angle of light. Separation of chromatic components of white light is exhibited in prisms as a result of refraction from different angular surfaces of a traditional prism. GRIN prisms, however, can utilize internal refractive index distributions to bend light, Figure 4.13, without the need for angled surface cuts.

As displayed in Figure 4.13, a planar GRIN optic, cut at an angle from a GRIN sheet, can act as a light deflector that is identical to a traditional prism. Additionally, a GRIN prism will deflect independent of external environmental conditions, that is, submersion in air versus water, stemming from the flat surfaces and optical power lying solely as a function of the intraprism nanolayered film based refractive index distribution.

A second nonimaging application of GRIN material involves redistributing light intensity profiles in beam-shaping applications [27]. Traditional lasers and light have a Gaussian intensity distribution, that is, a maximum intensity at the beam center quickly dissipates from the beam center to an edge. This significant variation in laser intensity is not ideal for applications such as laser cutting and welding, where the sharpness and time required for cutting/shaping material is highly dependent on the intensity of the laser beam. Conversion of the Gaussian laser beam intensity profile to a "flat top" shaped profile with a significantly lower difference between the beam edge and center intensity, is preferred in the laser cutting industry. The ability of GRIN optical material to utilize its bulk material to bend and redirect light, allows for the design and fabrication of just such a system (Figure 4.14). This beam shaper system utilizes two convex optics to redistribute the laser intensity from a normal Gaussian to a more desirable "flat top" distribution.

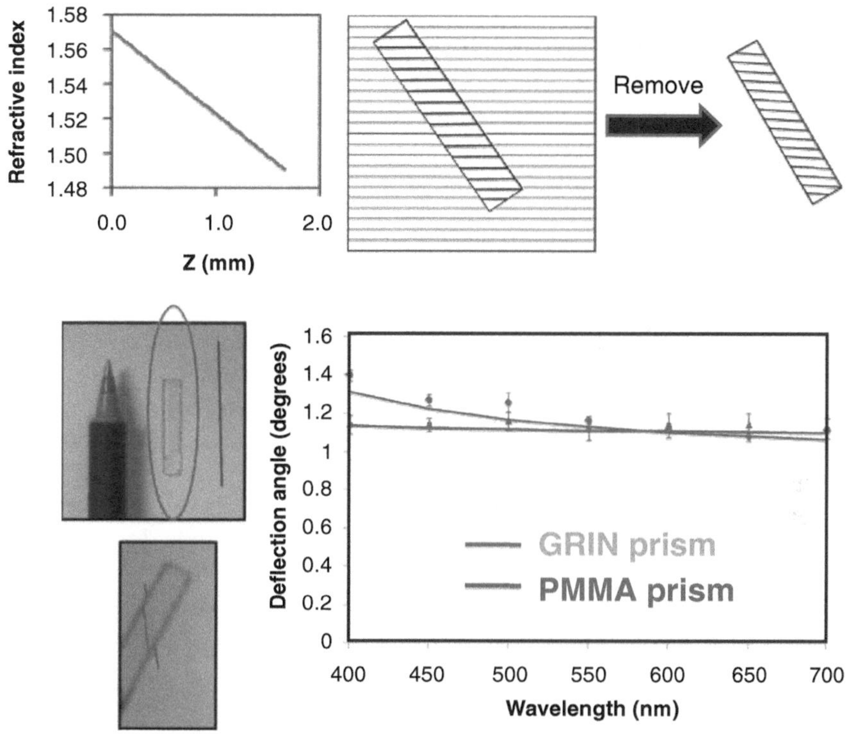

Figure 4.13 Nanolayer film enabled GRIN "flat prism."

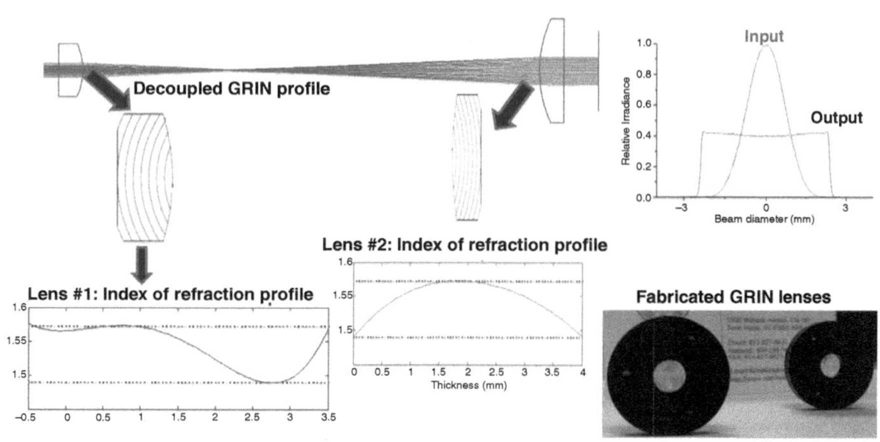

Figure 4.14 Nanolayer film enabled flat-top beam shaping GRIN lens system.

Whether the application is improving image quality, reducing weight, or redirecting light without reliance on external optic surface shape, the ability to produce nanolayered polymeric films with individual layer thicknesses < $\lambda/4$ of light, has enabled the fabrication of a class of GRIN optical materials, with unprecedented freedom and optical design flexibility to create new optical systems.

4.6 Summary

The micro- and nanolayer coextrusion process enables precise ordering of two and three material systems into periodic structures with a size scale to interact with light. Whether it is inducing a structural reflection in 100 nm to 1 μm thick layers to produce deeply colored or metallic looking Bragg crystal films without the use of pigments, or produce highly ordered sub-50 nm layers to trick the light to produce transparent, tailorable refractive index films, nanoscale level control, and ordering of polymer materials provides an optically relevant technology development platform for the identification and exploitation of unique physical phenomena in "ordinary" polymer materials. Although first realized in the 1960s, today's understanding and utilization of high levels of control in the coextrusion processing technique provides scientists, engineers, and physicists a highly developed technology platform to explore both diffractive and refractive advantages of controlling nanoscale 3D order in polymer materials and composites in size scales, which are relevant and interact with light. Modern applications to harness the utility of nanolayered optical films greatly surpass the historical commercial past of decorative only niche applications. This modern approach pushes interdisciplinary research in the academic and commercial research development environments; it will continue to realize and harness a unique process platform for nanolayer coextrusion and its unique capabilities, produce low cost mass producible products, and increase availability of processing equipment and accessible development lines. Realization of incorporating the nanolayered films into devices such as DBR lasers or discrete refractive index building blocks, in customized-lens design demonstrates some of these applications that are based on systematic development opportunities. As continued advances in inorganic particle, functionalized chemical molecules and dyes, and improved processing techniques push for even higher levels of control in the layering and ordering of materials, opportunities for utilizing nanolayer scale polymeric structure should continue to grow.

References

[1] T. Alfrey Jr., E.F. Gurnee, W.J. Schrenk, Polym. Eng. Sci. 9 (6) (1969) 400–404.
[2] M. Francon, Interscience Tracts, John Wiley and Sons, Inc, New York, 1963.
[3] A. Vasicek, Optics of Thin Films, North Holland Publishing Co, The Netherlands, 1960.
[4] T. Kazmierczak, H. Song, A. Hiltner, E. Baer, Macromol. Rapid Comm. 28 (2007) 2210–2216.
[5] M.F. Weber, C.A. Stover, L.R. Gilbert, T.J. Nevitt, A.J. Ouderkirk, Science 287 (5462) (2000) 2451–2456.
[6] M. Ponting, T.M. Burt, L.T.J. Korley, J. Andrews, A. Hiltner, E. Baer, Ind. Eng. Chem. Res. 49 (23) (2010) 12111–12118.
[7] The Mearl Corporation, US Patent No. 4,310,584, 1979.
[8] The Mearl Corporation, US Patent No. 5,008,143, 1987.
[9] W.R. Grace & Co., Cryovac Division, US Patent No. 4,643,943, 1985.
[10] The Dow Chemical Company, US Patent No. 5,149,578, 1989.
[11] The Mearl Corporation, US Patent No. 5,837,359, 1996.
[12] 3M Innovative Properties Company, US Patent No. 6,111,696, 2000.
[13] 3M Innovative Properties Company, US Patent No. 7,236,296, 2001.
[14] The Dow Chemical Company, US Patent No. 4,937,134, 1989.
[15] S. Armstrong, Z. Zhou, A. Hiltner, E. Baer, J. Appl. Polym. Sci. 113 (2009) 2150–2159.
[16] K. Singer, T. Kazmierczak, J. Lott, H. Song, Y. Wu, J. Andrews, E. Baer, A. Hiltner, C. Weder, Opt. Express 16 (14) (2008) 10358–10363.
[17] J.P. Dowling, M. Scalora, M.J. Bloemer, C.M. Bowden, J. Appl. Phy. 75 (4) (1994) 1896–1899.
[18] H. Song, K.D. Singer, J. Lott, Y. Wu, J. Zhou, J. Andrews, E. Baer, A. Hiltner, C. Weder, J. Mater. Chem. 40 (19) (2009) 7520–7524.
[19] R. Tangirala, E. Baer, A. Hiltner, C. Weder, Adv. Funct. Mater. 14 (6) (2004) 595–604.
[20] C. Ryan, C.W. Christenson, B. Valle, A. Saini, J. Lott, J. Johnson, D. Schiraldi, C. Weder, E. Baer, K.D. Singer, J. Shan, Adv. Mater. 24 (2012) 5222–5226.
[21] J.S. Shirk, M. Sandrock, D. Scribner, E. Fleet, R. Stroman, E. Baer, A. Hiltner, NRL Rev. 2006 (2006) 53–61.
[22] M. Ponting, A. Hiltner, E. Baer, Macromol. Symp. 294 (1) (2010) 19–32.
[23] S. Ji, M. Ponting, R. Lepkowicz, A. Rosenberg, R. Flynn, G. Beadie, E. Baer, Opt. Express 20 (24) (2012) 26746–26754.
[24] Y. Jin, H. Tai, A. Hiltner, E. Baer, J.S. Shirk, J. Appl. Polym. Sci. 103 (2007) 1834–1841.

[25] G. Beadie, J.S. Shirk, A. Rosenberg, P.A. Lane, E. Fleet, A.R. Kamdar, Y. Jin, M. Ponting, T. Kazmierczak, Y. Yang, A. Hiltner, E. Baer, Opt. Express 16 (15) (2008) 11540–11547.
[26] P. Meemon, J. Yao, K.S. Lee, K.P. Thompson, M. Ponting, E. Baer, J.P. Rolland, Sci. Rep. 3 (1709) (2013) 1–10.
[27] R. Zahreddine, R.S. Lepkowicz, R. M. Bunch, E. Baer, A. Hiltner, Proceedings of SPIE 7062 Laser Beam Shaping IX, Paper number 706214, 2008.

5 Dielectric and Electrical Properties of Multilayered Films

5.1 Introduction

Polymer, ceramic, and electrolytic capacitors are widely used in a variety of applications. Dielectric polymer film capacitors with graceful failure, are more advantageous than the ceramic and electrolytic capacitors for high-voltage/high-current pulsed power, power conditioning, and other applications [1]. Many other polymer film capacitor applications include consumer electronics, power systems, hybrid vehicles, and grid converters [2]. Commercially biaxially oriented polypropylene (BOPP) film is widely used for its low losses (tan $\delta \sim 0.0002$), and stable capacitance at room temperature. However, the low dielectric constant of 2.25 in BOPP films result in reduced energy density, close to 5 J/cm^3 at breakdown. In addition, BOPP also has significant DC conduction losses above 100°C, preventing its use at high temperatures [3]. Another commonly used polymer capacitor material, biaxially oriented poly(ethylene terephthalate) (PET) film, has a slightly higher energy density (6.0 J/cm^3 at a breakdown of 600 MV/m) and low dielectric loss (tan $\delta \sim 0.003$) [4]. Low energy densities of these commercial polymers lead to a large volume of capacitors in most of the applications. Development and scale-up of materials, with high dielectric constant and high breakdown strength, and with improved energy storage capabilities, is necessary to produce compact capacitors. Furthermore, polymer films with improved temperature capabilities and high energy density, and with low losses are also beneficial. This chapter discusses multilayered films with improved dielectric performance, which was achieved by combining polymers with contrasting dielectric properties.

The energy density (U_d) of the capacitor for a linear, ideal dielectric material, is defined by Equation (5.1) as,

$$U_d = \frac{1}{2}\varepsilon_o \varepsilon_r E_B^2 \tag{5.1}$$

where ε_o is the vacuum permittivity (8.85×10^{-12} C^2/Jm), ε_r is the relative permittivity (dielectric constant) of the material, and E_B is the applied electric field. It is obvious that increasing the dielectric constant and/or the breakdown strength increases the energy storage capability

Manufacturing and Novel Applications of Multilayer Polymer Films
http://dx.doi.org/10.1016/B978-0-323-37125-4.00005-8

of a dielectric material. Various approaches have been investigated to combine two or more components to achieve this synergistic effect. For example, composites prepared by combining high breakdown strength polymer, with high permittivity inorganic particles, such as barium titanate and titanium dioxide, improved the overall dielectric constant but negatively impacted the breakdown strength [5]. Many other nanoparticles are also investigated using this approach as well as by chemical modification, to improve their dispersion in the polymer composites. Particle agglomeration and scalability are some of the key issues associated with these approaches. Other polymers like polyvinylidine-*co*-trifluoroethylen-based terpolymers formed nanodomains, to improve the dielectric properties, but are commercially not viable [6]. Another approach of layered dielectric ceramic materials revealed enhanced dielectric strength and increased energy density [7]. The layers offered a barrier effect in the dielectrics increasing the breakdown strength, and also produced electrical treeing effect. The breakdown voltage and electrical trees in a quasi-uniform electric field for the layered composites were dependent upon several parameters such as the ratio of permittivities, polarization, homogeneity of the materials, position of the layer boundaries, conductivities of the barrier, space charge, etc. [8].

As discussed, the advantages of barrier layers and interfaces were extended to commercially scalable multilayered polymer films produced by combining different polymer materials. A high dielectric permittivity material such as polyvinylidene fluoride (PVDF) ($\varepsilon_r = 11–15$) was separated in a multilayered structure by polycarbonate (PC) layers with high-dielectric strength ($E_b > 600$ kV/mm, $\varepsilon_r = 3$). PVDF polymers have been investigated for piezoelectric and pyroelectric properties, for use in actuators, transducers, and heat sensors. However, it shows large dielectric hysteresis due to dipole switching and causes significant losses, which makes it an unattractive material for capacitor applications. In contrast, PC shows linear dielectric hysteresis response, with very small losses, and a low energy density. The synergistic effect of the two polymers namely high dielectric constant of PVDF and high breakdown strength of PC – was achieved in multilayered films with improved dielectric properties. Multiple examples of multilayered dielectric films with enhanced energy density, low dielectric loss using PVDF or its copolymers, a high electric breakdown polymer (e.g., PC and polyethylene terephthalate (PET)) have been reviewed in this chapter. Various parameters including the number of layers, layer thicknesses, film thicknesses, interphases, and orientation have been reviewed to understand the improved dielectric properties. Improved energy density implies compact capacitors for commercial applications.

In contrast to dielectric multilayered films, this chapter briefly reviews the examples of metal-filled multilayer films and the effect of layering on conductivity anisotropy. Metal-filled multilayer films have demonstrated anisotropic conductivity properties by improving the orientation of metal fillers in the coextruded structures.

5.2 Dielectric Films

5.2.1 Energy Density of PC/P(VDF–HFP) Multilayered Films

Multilayered films of PC and polyvinylidene difluoride-hexafluoropropylene copolymer, P(VDF–HFP), were investigated for their various dielectric properties. The films demonstrated improved energy density when the individual layer thicknesses were in the range of a few tens to a few hundred nanometers [9]. In this study, two sets of coextruded films with 32 and 256 layers of PC and P(VDF–HFP), respectively were produced in $[A/B]_n$ arrangement with varying compositions (reported as volume/volume ratio) as shown in Table 5.1. The overall film thicknesses were maintained at 12 μm in all the samples.

The relative permittivity, breakdown strength and energy density measurements were performed using different electrode geometries. The circular electrodes with 6.2 mm diameter were sputter coated on the film for capacitance measurement. The low field permittivity, ε_r, was calculated using $\varepsilon_r = c_p d / \varepsilon_o A$, where c_p is the capacitance, d is film thickness, and A is the area of the device. The breakdown strength of these films was measured using sphere-plane electrode geometry. The plane electrode of 1 × 1 in.^{2} and the spherical electrode of diameter 2 in. was used for this measurement. The permittivity values and the breakdown strength of

Table 5.1 PC/P(VDF–HFP) Composition in 12 μm, 32-Layered Films

PC/P(VDF–HFP) Composition (v/v)	Layer Thickness (nm)
0/100	0/750
10/90	75/675
30/70	225/525
50/50	375/375
70/30	525/225
90/10	675/75
100/0	750/0

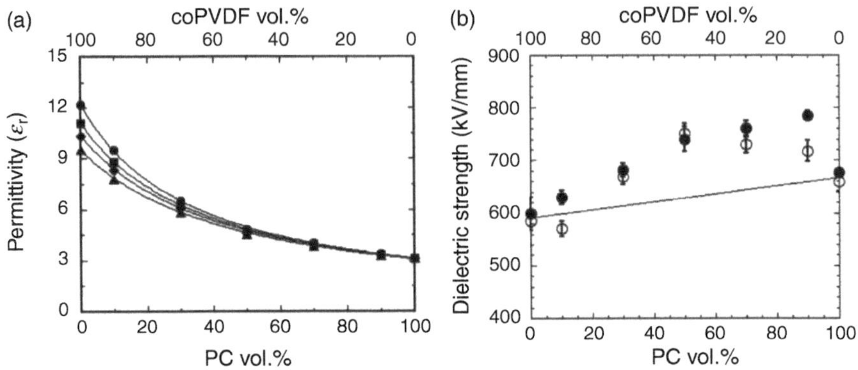

Figure 5.1 (a) Permittivity of 32-layered films as a function of volume concentration, measured at 100 Hz circles, 1 kHz squares, 10 kHz diamonds, and 100 kHz triangles. Solid lines represent permittivity calculated with Equation (5.2), (b) average dielectric strength of 32-layered solid circles and 256-layered open circles PC/P(VDF–HFP) (coPVDF) films as a function of volume concentration. The solid line represents the expected dielectric strength based on a weighted average.

32-layered films, as a function of PC and PVDF compositions, are shown in Figure 5.1. The permittivity of layered films increased with increasing concentration of P(VDF–HFP) and fitted into the layered effective medium or capacitors in series model as given in Equation (5.2) [10]

$$\frac{1}{\varepsilon_r} = \frac{\phi_1}{\varepsilon_{r1}} + \frac{\phi_2}{\varepsilon_{r2}} \tag{5.2}$$

where ϕ_1 and ϕ_2 are the volume fractions and ε_{r1} and ε_{r2} are the permittivity values of the two components. The measured permittivity values at different frequencies were within the reasonable range as calculated from the PC and P(VDF–HFP) control permittivity values as shown in Figure 5.1a. Unlike permittivity properties, a significant increase in the breakdown strength of layered composites as compared to the breakdown strength of PC or P(VDF–HFP)P(VDF–HFP) controls was unique to multilayered films, as shown in Figure 5.1b. The maximum breakdown strength was observed in 70/30 PC/P(VDF–HFP) composition for a 32-layered film. As PC and P(VDF–HFP) are immiscible, the comparable blends of the two components showed reduced breakdown strength due large amounts of interfaces that are parallel to the applied field. Furthermore, weighted average model and series model obtained by ideal circuit analysis, estimated the lower breakdown strength for multilayers than measured values. Increased breakdown strength also resulted in increased energy density according to Equation (5.1). From breakdown voltage of 1000 kV/mm and

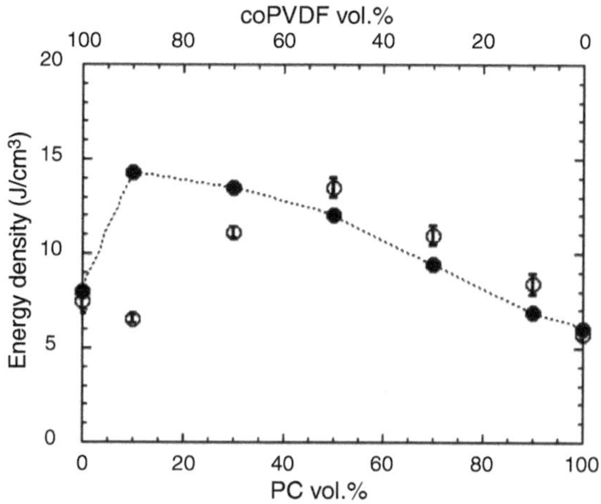

Figure 5.2 Maximum projected filled circles and measured open circles energy density of 256-layered PC/P(VDF–HFP) (coPVDF) films as a function of volume concentration. The projected value for the P(VDF–HFP) control film was also determined.

permittivity of 4.0, the energy density was calculated as high as 17 J/cm^3. Measured energy density values using charge/discharge circuit are shown in Figure 5.2. Energy density values for the 50/50 PC/P(VDF–HFP) compositions were measured as 14 J/cm^3, which was significantly higher than the PC or P(VDF–HFP) controls and close to the calculated values.

5.2.2 Breakdown Failure Analysis

To investigate the increased breakdown strength, analysis of the breakdown failure sites in the multilayered films was performed and compared with breakdown characteristics of the monolith controls and PC/PVDF blends [11]. The breakdown measurements were performed using needle–plane electrodes, with a 20 ± 5 μm diameter needle electrode tip and 20 × 51 × 2 mm brass plate plane electrode. Subsequent optical microscopy of the failure sites in the multilayered films revealed distinct tree-like patterns surrounding the main breakdown hole, while the monolithic control films showed single breakdown holes. The differences in these breakdown characteristics are shown by comparing 50/50 PC/P(VDF–HFP) films with PC control as shown in Figure 5.3a,b. It must be noted that a positive needle polarity was required to achieve treeing patterns. The slower electrical mobility of positive charge caused the accumulation of charge carriers resulting in high local electric fields at the interfaces and eventual interfacial failure, which is observed as a treeing phenomenon. The formation and

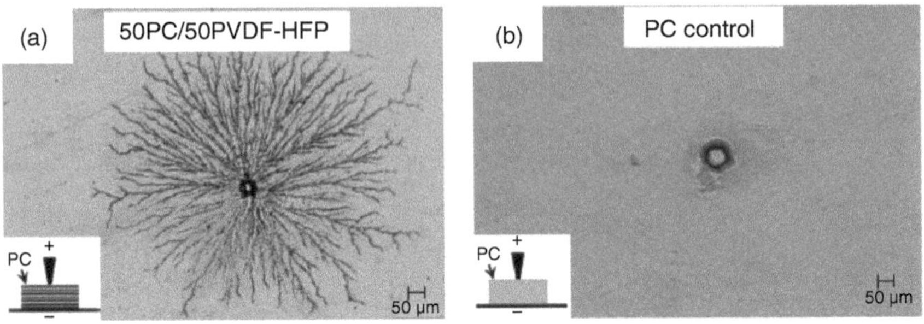

Figure 5.3 Optical micrograph damage zone images of 12 μm thick films subjected to 950 kV/mm field. Image (a) shows a 32-layered film, positive needle polarity, needle adjacent to PC and (b) shows PC single layer control, positive needle polarity.

evolution of treeing phenomena was investigated by varying the electric field from 450 kV/mm to 950 kV/mm in 12 μm thick PC/P(VDF–HFP) film with 50/50 composition. The films did not show any treeing pattern up to a field of 450 kV/mm and the corresponding current versus time plot also did not show any differences in input charge and discharge. With further increase in applied electric field, the size of treeing pattern increased almost linearly with the electric field, except at very high electric fields. Formation of treeing patterns without any breakdown hole at 550 kV/mm was associated with reduced output charge suggesting energy loss was associated with treeing formation. Further increase in the treeing size as a function of electric field correlated with the increased energy loss. The scanning electron microscopy (SEM) imaging revealed distinct damage of the layered films surface adjacent to the needle electrode and absence of treeing pattern on the other side of the film with plane electrode as shown in Figure 5.4. It was clear that the damage to these layered films was limited only to the first few layers of the film across the layer interfaces, in a divergent electric field.

The breakdown failure in 50/50 PC/P(VDF–HFP) samples was further analyzed using focused ion beam (FIB) milling and SEM techniques. This advanced approach of FIB milling removed submicrometer thickness sections from the materials with high precision [12]. SEM image of the failure sites in Figure 5.5 showed changes to the internal layered structure after breakdown event. Delaminated layers and voids can be seen in Figure 5.5b, which showed FIB milled specimen beyond the breakdown hole (pinhole). Spatial resolution of the multilayer structure in a breakdown area was carried out by systematically milling the samples at different distances from the breakdown hole, and imaging the sample cross-section.

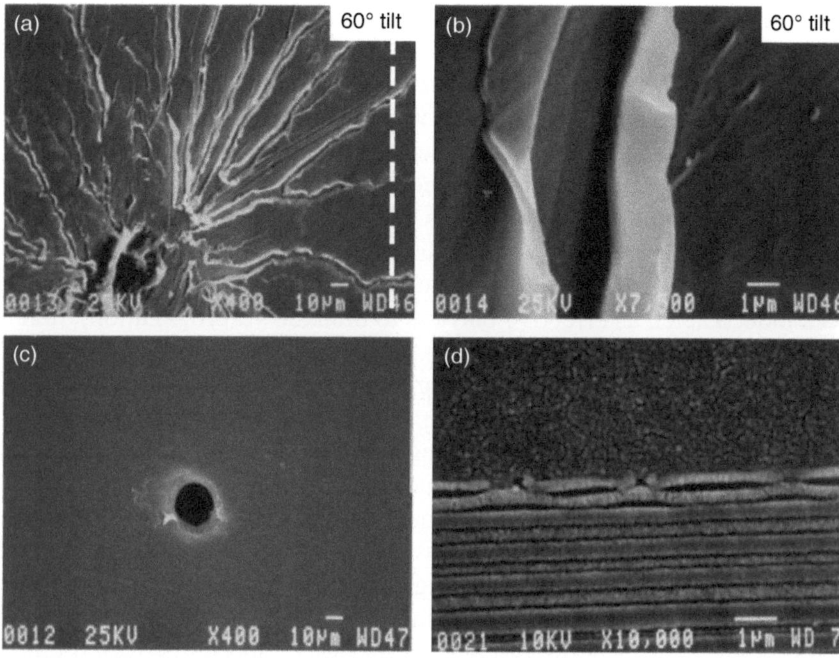

Figure 5.4 SEM micrographs of treeing damage of 32 layer, 12 μm, PC/P(VDF–HFP) (50/50) films with PC adjacent to the needle and subjected to a 1000 kV/mm field for 0.2 s, with a positively polarized needle electrode. Images (a) and (b) are views from needle's electrode side of film. Image (c) is a view from plane electrode side of film. Image (d) is the cross-section of treeing damage corresponding to the dotted line in (a).

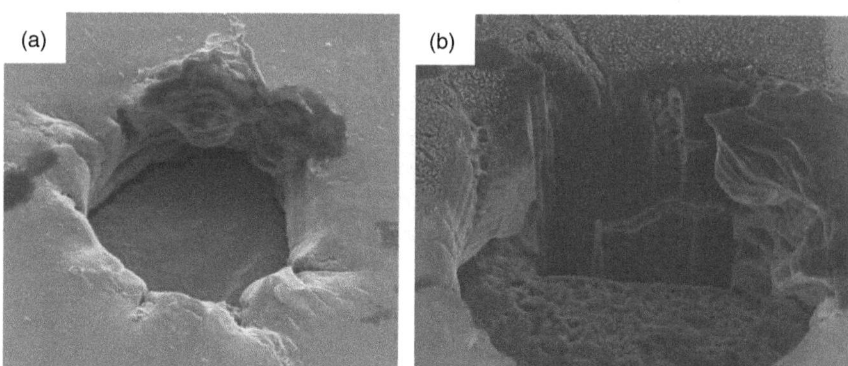

Figure 5.5 (a) SEM micrograph of a breakdown pinhole (d = 40 μm) in a 32-layered 50/50 PC/P(VDF–HFP) film. (b) Cross-section of the film at 2 μm into the film relative to the pinhole sidewall. This image was obtained after a 90° clockwise rotation from the orientation shown in (a).

Formation of isolated small voids at 35 μm distance from the breakdown hole and larger voids at layer interfaces at 14 μm distance from the breakdown hole were observed. Extensive layer delamination across multiple layers and significantly larger voids were visible at 7.5 μm distance from the breakdown hole. It is evident that the breakdown hole is associated with damages along the layer interfaces, several micrometers from the pinhole site across the film thicknesses with P(VDF–HFP) layer in contact with the high-voltage tip. When PC layer was in contact with high voltage tip, the damage was limited to surface and only a few interfaces. The interfaces indeed acted as effective barriers to breakdown propagation, thereby increasing the breakdown strength.

To further investigate the treeing breakdown phenomena, effects of layer and film thicknesses on treeing patterns were studied independently [11]. Multilayered films with varying number of layers ranging through 32, 64, 128, and 256 and film thicknesses in the range of 3–15 μm were coextruded. The breakdown phenomena and treeing patterns were investigated using needle–plane electrodes as a function of electric field. Measured treeing onset fields and treeing patterns were independent of layer thicknesses. More specifically, treeing was observed in film thicknesses greater than 6 μm. The input charge, discharge, and unrecovered charge calculated from current versus time plot, did not show any correlation with the layer thickness. Interestingly, the input charge increased linearly with electric field in the layered films. Upon treeing initiation in films thicker than 6 μm, the slope of input charge versus electric field increased rapidly. In contrast, films thinner than 6 μm showed a linear increase in input charge, up to the breakdown event without any treeing patterns. The data showed a minimum of 30–35 nC input charge required for treeing. This charge was identified as a critical input charge (Q_c), below which the input charge was directly proportional to the applied field and the proportionality constant increased with increasing film thickness. Increased proportionality constant translated into lower electric field requirement in thicker films for treeing onset. Such critical charge requirement was never met when the film thicknesses was less than 6 μm thick. The maximum breakdown strength occurs above Q_c as the treeing formation impedes the breakdown process in the layered films. A schematic correlating input charge, applied electric field, film thickness, breakdown strength, and treeing formation is depicted in Figure 5.6.

5.2.3 Breakdown Progression

The breakdown progression events and the long-term reliability of the dielectric films were tested using a dielectric lifetime test [13]. Earlier used

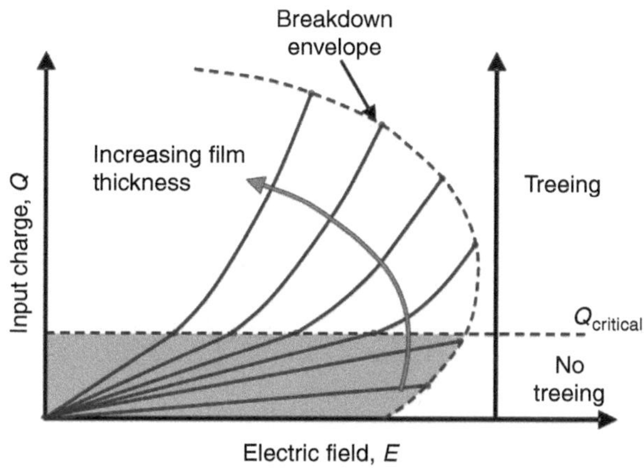

Figure 5.6 Diagram of high field behavior. Definition of the different regions: the critical charge, Q_c, separates the recoverable and treeing regimes; below Q_c the input charge is proportional to electric field; above Q_c, a disproportional amount of input charge is needed, a part of which is expended on treeing formation; the maximum dielectric strength on the breakdown envelope occurs with an input charge above the treeing threshold (Q_c), which shows that tree formation impedes the breakdown process.

needle–plane electrode configurations created a divergent field in the films resulting in treeing patterns [14]. Such ramped and pulsed voltage techniques are useful in assessing the short-term dielectric properties. However, commercial polymer film capacitors are typically evaluated using a plane–plane electrode configuration as it resembles the commercial testing and also generates a uniform field in the sample. For lifetime test, a plane–plane electrode configuration with a test area of a 1×1 cm^2 with gold-coated electrodes, was used for testing under a constant DC voltage. The applied field was maintained at 320 kV/mm and the breakdown events were recorded by measuring the electric current as a function of time. The individual breakdown events randomly occurred in the layered films and corresponded to different current spikes on the current versus time plot.

To understand the breakdown fracture characteristics, the breakdown fracture progression in a 32-layered 70/30 PC/P(VDF–HFP) film was analyzed using film breakdown images and recording capacitance values during the dielectric lifetime test. Images of breakdown progression with time are shown in Figure 5.7. These optical microscopy images showed random breakdown events in the film during the lifetime test. The number of breakdown holes increased and film capacitance decreased with time. Subsequently the breakdown sites interconnected above 20,000 s test time. Further breakdown occurred along the electrode edge with film

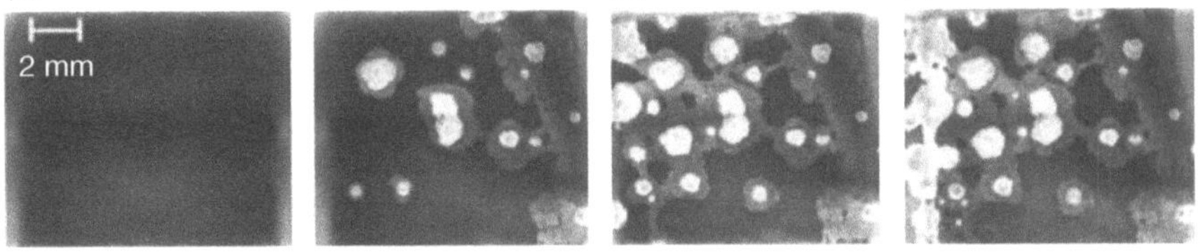

Figure 5.7 Breakdown progression of 32-layered 70/30 PC/P(VDF–HFP) with a thickness of 8 μm under 320 MV/m. The images were taken at different times. The times were approximately calculated from the current versus time plot as 0, 1500, 5000–6000 and 11,000 s, for images a, b, c, and d, respectively. The plot is published in Ref. [13].

capacitance reaching 0.05 nF. Based on the formation of breakdown holes, it is evident that the breakdown phenomena occurred in different stages as described by the capacitance change during this test. The capacitance measurements were performed by fitting the exponential decay of the charging current spikes showed inverse S-shaped plots as shown in Figure 5.8. The inverse S-shape plots indicate three breakdown regimes for multilayered films, also confirmed by two capacitance measurement techniques – dielectric thermal analyzer (DETA) and capacitance calculations through effective capacitance area. All the approaches yielded inverse S-shape of capacitance change with increasing test time, albeit with some differences in the absolute values. Nonetheless, the three regimes defined various breakdown characteristics in the multilayers. Slow capacitance decrease in

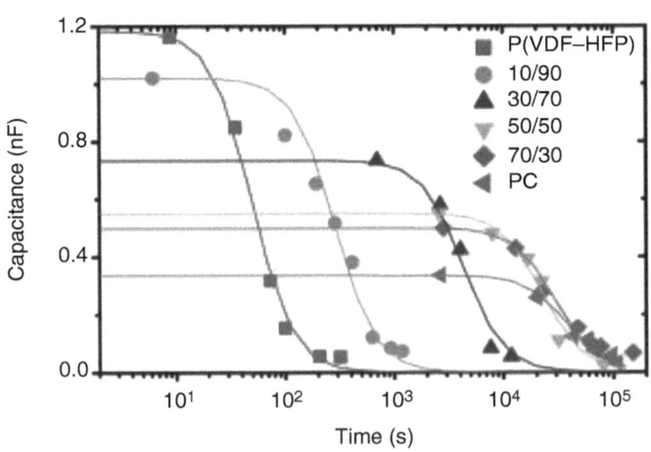

Figure 5.8 Film capacitance (from fitting of the charge current peak) dependence on the testing time for 32-layered PC/P(VDF –HFP) with different film compositions. The total film thickness is kept at 8 μm and the applied field is 320 MV/m.

the first regime was attributed to random, isolated breakdown events. The second regime with faster change in capacitance, resulted in interconnected breakdowns, while the capacitance leveled off in the last regime due to wearing-out, as most of the breakdown occurred along the electrode's edge. The slopes shifted to longer times with increasing PC content, suggesting shifts in the breakdown regimes to higher lifetimes.

5.2.4 Enhanced Dielectric Lifetime

PC/P(VDF–HFP) films (32 layers, 8 μm thickness) were investigated to evaluate the impact of composition (relative layer thicknesses) on the dielectric lifetime. The current profile and the damage sites in the films are shown in Figure 5.9 [13]. The multiple current spikes in Figure 5.9a represent the breakdown events in films. For P(VDF–HFP) control, majority of the breakdown events occurred at very short times of 10^2 s or less, while PC control showed breakdown events close to 10^5 s. Interestingly, the layered films showed increased dielectric lifetime with increasing PC concentration. When PC layers were very thin (50 nm) in 10/90 PC/PVDF composition, the leakage current in the layered films was significantly above zero and similar to P(VDF–HFP) control. It was attributed to poor insulating resistance of P(VDF–HFP) layers. Increasing the PC concentration to 30% (PC layer thickness of 150 nm) showed an enhancement of more than one order of magnitude in dielectric lifetime as compared P(VDF–HFP) control films. Further increase in PC content showed lifetimes between 10^4 s and 10^5 s. The observations suggested a critical PC layer thickness requirement to enhance the dielectric lifetimes.

The analysis of breakdown sites was performed to correlate charge/discharge events with the current spikes. The damage sites in Figure 5.9b showed that the number of breakdown holes were significantly less than the number of current spikes in a current profile. As discussed, the main breakdown hole formation was associated with several interfacial delamination events and void formations. Therefore, different current spikes must represent the breakdown holes as well as interfacial discharge events. The intensity of current spikes was correlated with the breakdown events to separate these two breakdown phenomena. It was observed that current spikes formed at higher currents (>7.5 μA) correlated well with a number of breakdown holes in the film. The remaining low amplitude current spikes were attributed to surface discharge or expansion of the existing breakdown holes.

Further analysis of the failure data of the films revealed a bimodal Weibull distribution corresponding to two different breakdown regimes. The two slopes in the plots as shown in Figure 5.10a represent the isolated breakdown and interconnecting breakdown events discussed in a

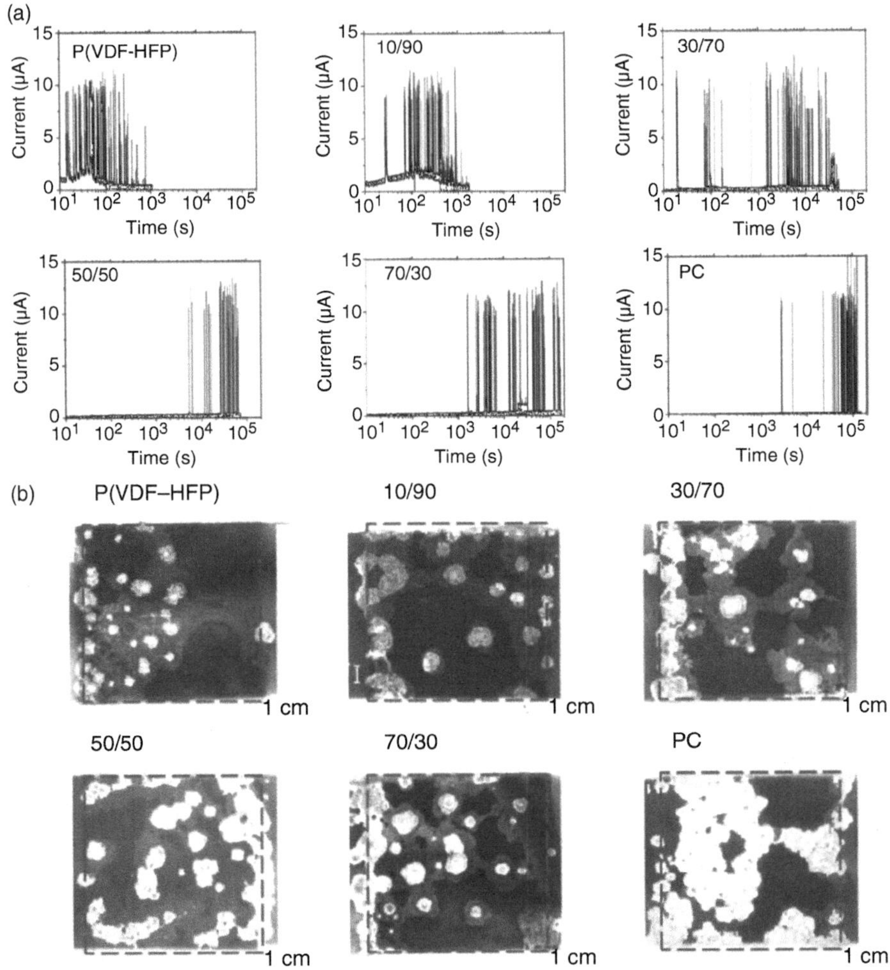

Figure 5.9 (a) Effect of film composition on the current profile of the 32-layered PC/P(VDF–HFP) films during dielectric lifetime tests. (b) Effect of film composition on the damage sites of the 32-layered PC/P(VDF–HFP) films. The total film thickness is kept at 8 μm and the applied field is 320 MV/m.

breakdown progression. The change in these slopes correlated well with the capacitance change associated with different breakdown events. The failure data for current spikes above 7.5 μA (only breakdown holes) was analyzed using a two-parameter Weibull distribution function given as follows:

$$P_{\mathrm{F}}(t) = 1 - \exp\left[-\left(\frac{t}{\tau}\right)^{\beta}\right] \tag{5.3}$$

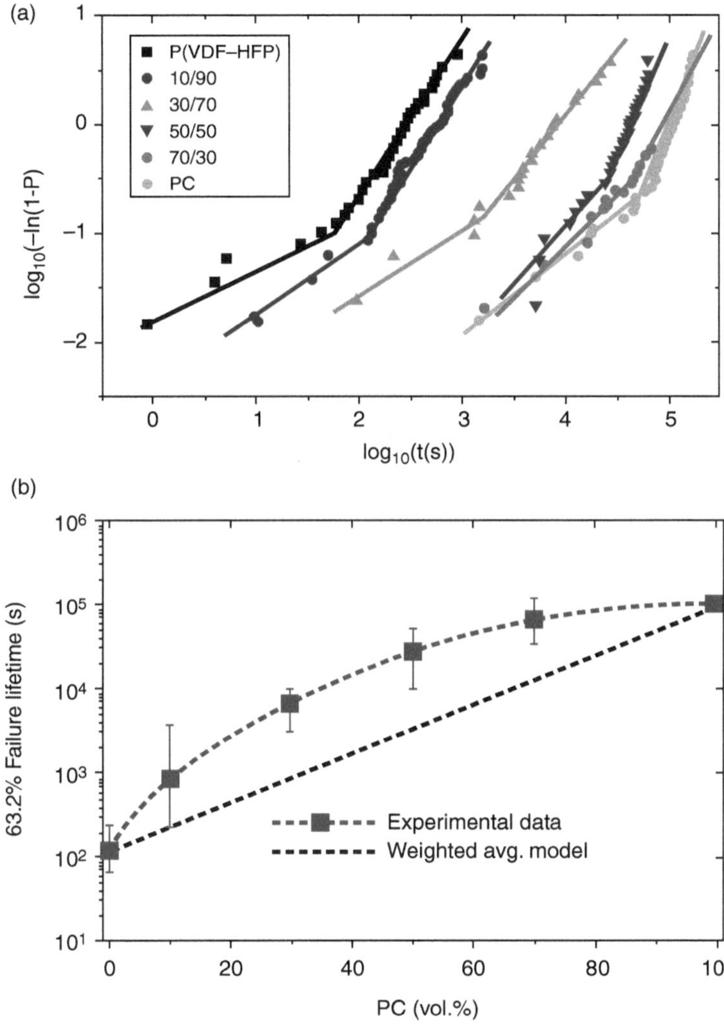

Figure 5.10 (a) Weibull analysis of the breakdown data of 32-layered PC/P(VDF–HFP) films with various compositions; (b) characteristic lifetime of the 32-layered PC/P(VDF–HFP) films from Weibull analysis. The total film thickness is kept at 8 lm and the applied field is 320 MV/m.

where $P_F(t)$ is a cumulative probability of failure, t is the characteristic dielectric lifetime representing 63.2% failure probability. Figure 5.10b shows the experimental failure lifetime in layered films, and it was significantly higher data than the weighted average model prediction. The layered films with 70/30, 50/50, and 30/70 compositions showed at least

Table 5.2 **PC/P(VDF–HFP) Film Sample Details for 50/50 Composition (10 μm) Thickness**

HFP layer thickness (nm)	5000	2500	1250	630	310	160	80	40	20
No. of layers	3	5	9	17	33	65	129	257	513

an order of magnitude improvement in failure lifetime over the weighted average values. Comparable blends showed 63.2% failure lifetimes in the range of 10–100 s, which were at least two orders of magnitude smaller than the layered films' failure lifetimes. Interestingly, the failure lifetimes in 70/30 PC/P(VDF–HFP) films did not change significantly when the number of layers increased from 32 to 256. However, both 32- and 256-layered films showed significant improvements as compared to 2-layered films.

The effect of total number of layers and layer thicknesses on the dielectric lifetime was further investigated in PC/P(VDF–HFP) films (50/50 composition, 10 μm thickness) with varying number of layers and layer thicknesses as shown in Table 5.2 [15]. The number of layers varied from 3 to 513 and corresponding layer thicknesses changed from 5 μm to 20 nm. The dielectric lifetime results as a function of layer thickness and number of layers are shown in Figure 5.11. PC control, P(VDF–HFP) control and a 50/50 blend sample was also included for the comparison. Melt-blended PC and P(VDF–HFP) film produced phase separated blends resulting in early failure. The presence of interfaces parallel to the electric field enhanced the dielectric lifetime in multilayered films as compared to P(VDF–HFP) control and the blend film. The dielectric lifetime increased with increasing number of layers up to 65 layers with 160 nm layer thicknesses, which showed the maximum improvement in the dielectric lifetime. Further increase in the number of layers reduced the dielectric lifetime. Analysis of the breakdown sites showed layer thickness dependent interfacial delamination and void formation. The breakdown zones in thicker layers (nine-layered film) extended as far as 50 μm away from the main breakdown site.

The effect of layer thickness and number of layers on the breakdown fracture properties is summarized in Figure 5.12. Three types of failure characteristics were observed as a function of layer thickness. Films with layer thicknesses ranging from 5 μm to 0.63 μm (3 and 17 layers) showed formation of noninteracting crazes within the P(VDF–HFP) layers. As the crazes remained confined to P(VDF–HFP) layers, the height of the crazes

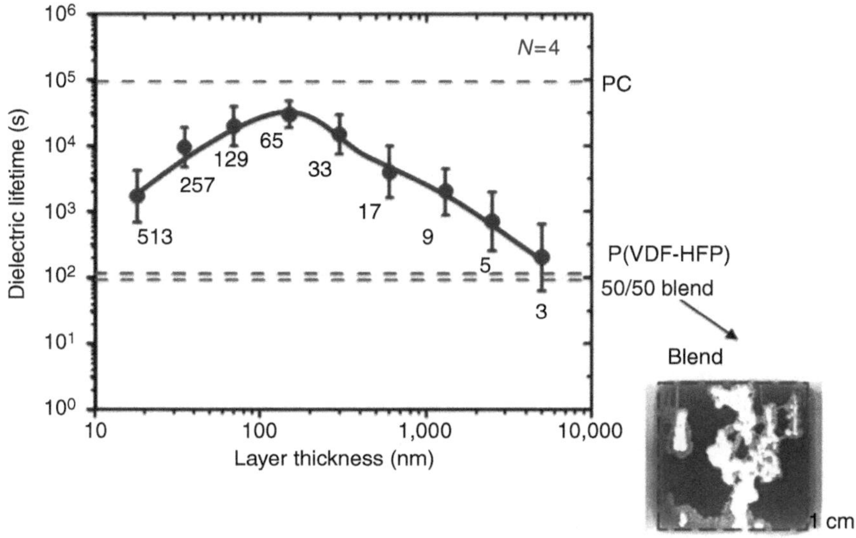

Figure 5.11 Dielectric lifetime as a function of layer thickness for PC/ P(VDF–HFP) films with various numbers of layers. The PC, P(VDF–HFP), and 50/50 blend were also included and indicated as the dashed lines. The numbers next to the dots are the layer numbers of the film tested. The optical microscope image shows all damage holes on the 50/50 PC/PVDF–HPF blend film after dielectric lifetime test.

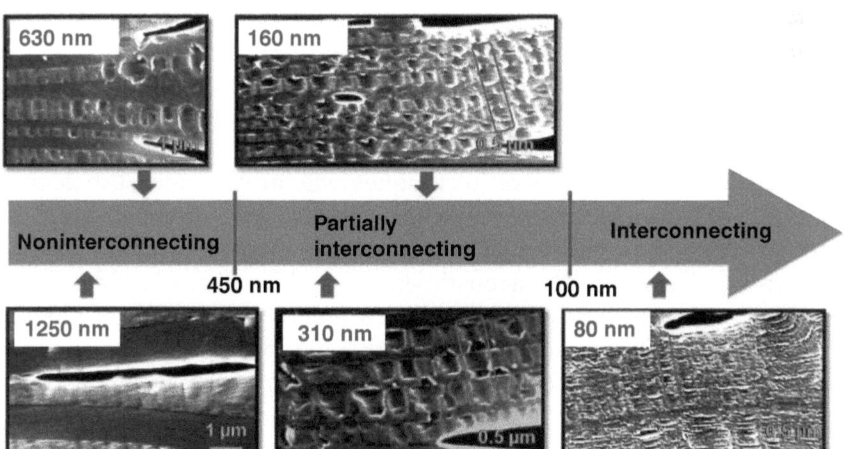

Figure 5.12 Representative images showing the effect of layer thickness on the type of crazing next to the breakdown hole after dielectric breakdown. Layer thicknesses for the samples are indicated on the top left corner of each image. The total film thickness is kept at 10 μm and the applied field is 320 MV/m.

was same as the layer thicknesses, while the width of the crazes was approximately 400 nm and contained 50 nm stretched fibrils within the craze. Interestingly, the PC layers did not show any morphological changes and remained continuous. Layer delamination was also occasionally observed in these films indicating weaker adhesion of PC and P(VDF–HFP) polymers. Further reduction in layer thicknesses (100–450 nm) showed partially interconnecting crazes through two to three layers across the film thickness. Layer thicknesses below 100 nm showed craze arrays with craze penetration through several PC layers. The craze arrays in the thinner PC layers were not effective in preventing breakdown propagation through the films, as evident by reduced dielectric lifetime when the number of layers changed from 65 to 513. In summary, multilayered films with layer thicknesses above 100 nm showed single crazes while interconnecting crazes or craze arrays throughout the entire sample were common in layer thinner than 100 nm.

The changes in crazing phenomena also affected the craze density. A craze density of 20 crazes per 10 μm^2 in 1.3 μm layers (nine-layered film) increased substantially to 160 crazes per 10 μm^2 in 160 nm layers (65-layered film) followed by decreased craze density with further reduction in layer thickness. Majority of the crazes in the films with layer thickness above 100 nm (65 layers and below) were single crazes. The increased crazing density correlated with increased dielectric lifetime, as long as the films formed single crazes. In thinner layers, crazing propagation through thin PC layers resulted into craze arrays and early failure. It was obvious from these observations that a minimum critical PC layer thickness was required to reduce craze propagation in the films.

It is clear from the breakdown image analysis that breakdown fracture consisted of two events – formation of main breakdown hole and interfacial fracture. To differentiate between the different breakdown events, an approach of acoustic emission techniques was used to measure acoustic amplitude of the breakdown events. The acoustic measurement set-up was used in conjunction with the dielectric lifetime measurements to correlate the current spikes with the acoustic emission. The acoustic signals were amplified and band pass filtered within a predefined frequency range. A threshold of 60 dB was selected to remove any background noise during the test. An example of electric current profile along with the acoustic amplitude data in Figure 5.13a,b shows good correlation between the acoustic hits and the current spikes. Interestingly, the intensity of acoustic hits changed for different current spikes. The intensity of the acoustic hits also changed dramatically with layer thicknesses and number of layers. In fact, the input energy charge and acoustic amplitude correlation showed that the highest or initial acoustic hits had the maximum amplitude, as shown

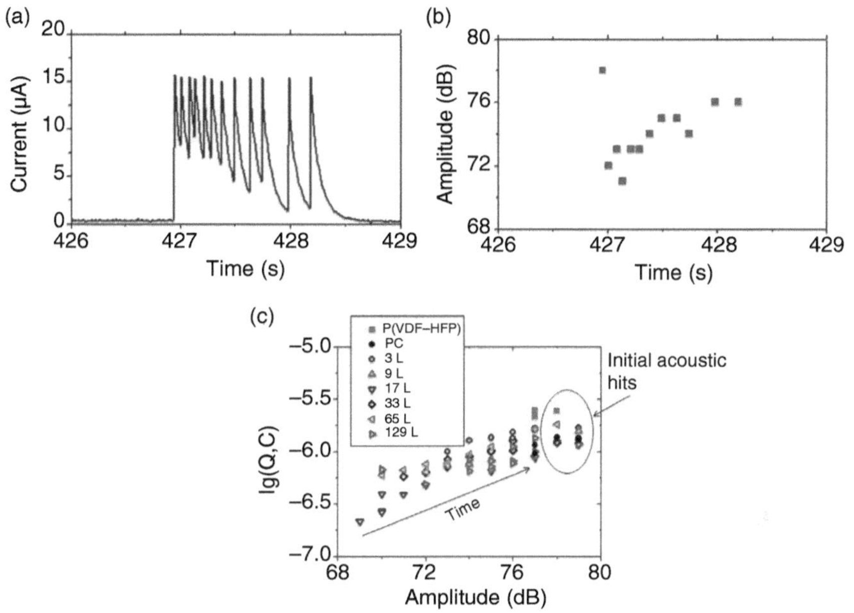

Figure 5.13 (a) Electric current profile and (b) acoustic amplitude data for the breakdown event of the 65-layered 50/50 PC/P(VDF–HFP) film; (c) correlation between the acoustic amplitude and the input charge obtained from current profile in the 50/50 PC/P(VDF–HFP) system with various numbers of layers. The total film thickness is kept at 10 μm and the applied field is 320 MV/m.

in Figure 5.13c. Subsequent hits showed reduced amplitude irrespective of layer thickness. The maximum acoustic amplitude was observed in PC and PVDF monolayer controls. As these control films showed single breakdown hole, the high amplitude acoustic hits must represent discharge from top to bottom electrode. In contrast, the low amplitude events arose from the internal discharge along layer interfaces. The number of acoustic hits increased with increasing number of interfaces suggesting easy discharge along the interfaces.

Based on these observations, the breakdown mechanism for the layered film was proposed and schematically shown in Figure 5.14 [15]. When the electric field is applied, the hot spot along the conductive pathways is formed across the film resulting into microsecond discharge [16]. The quick energy discharge created localized hot spots and also caused pressure build-up within the film. As the pressure increased to a level that is more than the mechanical strength of the film, film discharge resulting into a breakdown hole is formed. Evaporation of the metal electrodes at the breakdown site interrupts the breakdown event and conductive pathways

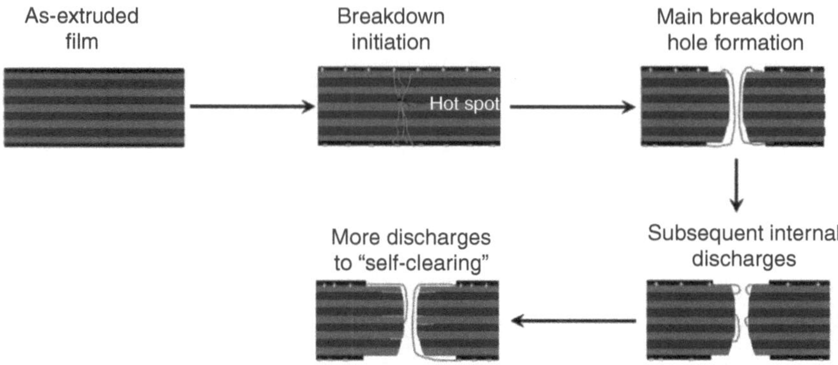

Figure 5.14 Schematic representation of the time evolution of a breakdown event. Black, light gray [red in the web version], and dark gray [blue in the web version] colors represent electrodes, PC layers, and P(VDF–HFP) layers, respectively. Solid gray lines [green lines in the web version] represent electric current.

are decomposed rapidly preventing catastrophic failure of the film. Subsequent charging of the capacitor film represented by current spikes is known as self-healing or self-clearing [17]. In addition, as evident from the SEM images, layer thickness dependent microcavitation, crazing, and interfacial delamination were also associated with the breakdown event, as indicated by low amplitude acoustic events. Eventual evaporation of the electrodes prevents breakdown and catastrophic failure as the conductive pathways disappear at the breakdown site. Although creation of interfaces inhibits the damage propagation through the films, a critical layer thickness is required for increased lifetime as thinner layers can fail by formation of craze arrays and by rapid craze propagation through the film.

5.2.5 Polarization in Multilayer Films

P(VDF–HFP)-based multilayer films demonstrated improved energy density, breakdown strength and dielectric lifetimes. Layer interfaces and relative layer thicknesses were critical in improving these properties. It is necessary to understand the dielectric polarization and losses in the multilayered films in developing applications [18]. The polarizations responsible for dielectric losses in polymers are identified as electronic, vibrational (atomic), orientational (dipolar), ionic, and interfacial polarizations are observed in polymers. The electronic and vibrational polarization losses occur in infrared and optical frequency range, also known as resonance regime. In contrast, orientation related, ionic, and interfacial polarizations occurred over a broad range of frequencies ($<10^{-9}$–10^{5} Hz), also known as relaxation regime. PVDF-based ferroelectric polymers exhibit orientational polarization due to dipolar switching under applied electric field resulting into hysteresis. In addition, the presence of ions in the raw material (from the suspension polymerization process) can result into ionic polarization, while the charge carriers accumulated at the polymer–polymer interface in the multilayered films may cause interfacial polarization. As all these polarizations contribute to losses, it was necessary to understand the effect of layering on these polarizations. The high field hysteresis and the polarizations were investigated in PVDF homopolymer-based multilayered films. Multilayered films of PC and PVDF showed increased energy density and significantly reduced dielectric losses, similar to PC/P(VDF–HFP) multilayer films.

5.2.5.1 High-Field Hysteresis

The high-field polarization hysteresis in PC/PVDF films as a function of number of layers and film thicknesses was investigated. The layer thicknesses in the films ranged from 6 μm to 25 nm as the number of layers changed from 2 to 256. Film sample details are shown in Table 5.3 [19]. The AFM images of the film cross-section in Figure 5.15 confirmed that the measured layer thicknesses were close to the target nominal layer thicknesses. Electric displacement-electric field (D-E) hysteresis measurements were conducted by applying voltage in a bipolar triangular waveform at

Table 5.3 Details of PC/PVDF (50/50, v/v) Multilayer Films and the Corresponding Thermal Properties

PC/PVDF Composition (v/v)	No. of Layers	Film Thickness (μm)	PVDF Layer Thickness (nm)	PVDF Crystallinity (%)	T_m (°C)	T_c (°C)
PVDF	Control	–	12	51	171	131
PC	Control	–	12	–	T_g: 150	–
50/50	2	12	6000	48	171	132
50/50	2	6	3000	50	171	131
50/50	8	12	1500	46	171	131
50/50	8	6	750	45	171	132
50/50	32	12	380	45	171	130
50/50	32	6	187	43	172	132
50/50	256	12	50	42	171	128
50/50	256	6	25	42	171	129

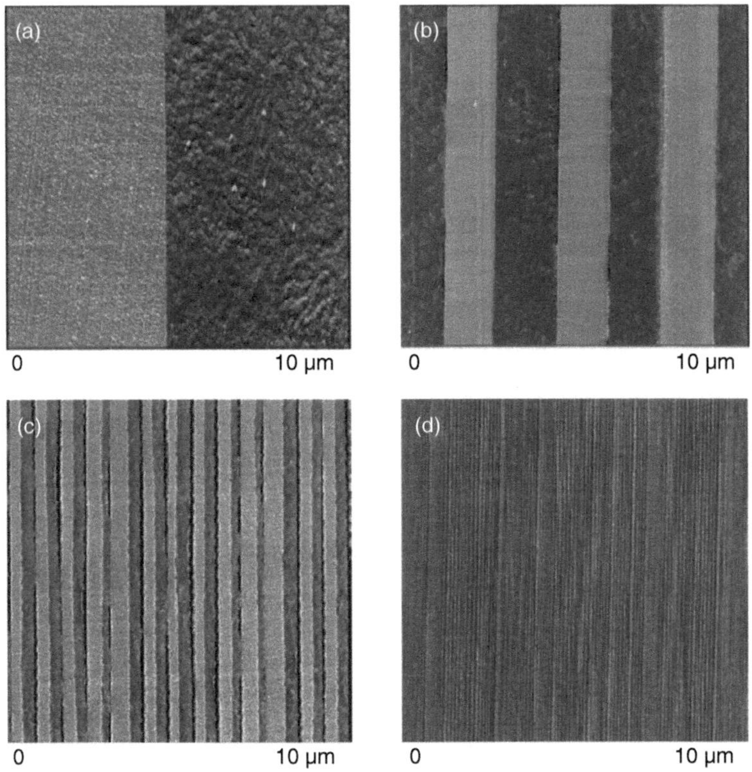

Figure 5.15 AFM phase images of 50/50 PC/PVDF layered film cross-sections of films. With (a) 2 layers (12 μm/6000 nm); (b) 8 layers (12 μm/1500 nm); (c) 32 layers (12 μm/375 nm); and (d) 256 layers (12 μm/50 nm). The light and dark colored layers are PC and PVDF, respectively.

1 Hz on a 1 cm diameter film area sandwiched between two electrodes. Hysteresis properties of the multilayered films were compared with the monolayer control films. Linear dielectric material (PC) exhibited low losses and the ferroelectric material control (PVDF) showed a large hysteresis loop with dipole saturation at high applied fields, as shown in Figure 5.16 [20]. Interestingly, the multilayered films did not show any dipole saturation for PVDF layers at comparable applied fields. The films showed linear D-E loops with higher loop area than PC and without any apparent dipole saturation of PVDF. As the number of the layers increased from 2 to 256, substantial decrease in the hysteresis loop area was evident with decreasing layer thickness. The other PC/PVDF film samples with offset compositions were also characterized by D-E hysteresis measurement. Reduced hysteresis was observed when the PVDF layers were thinner than

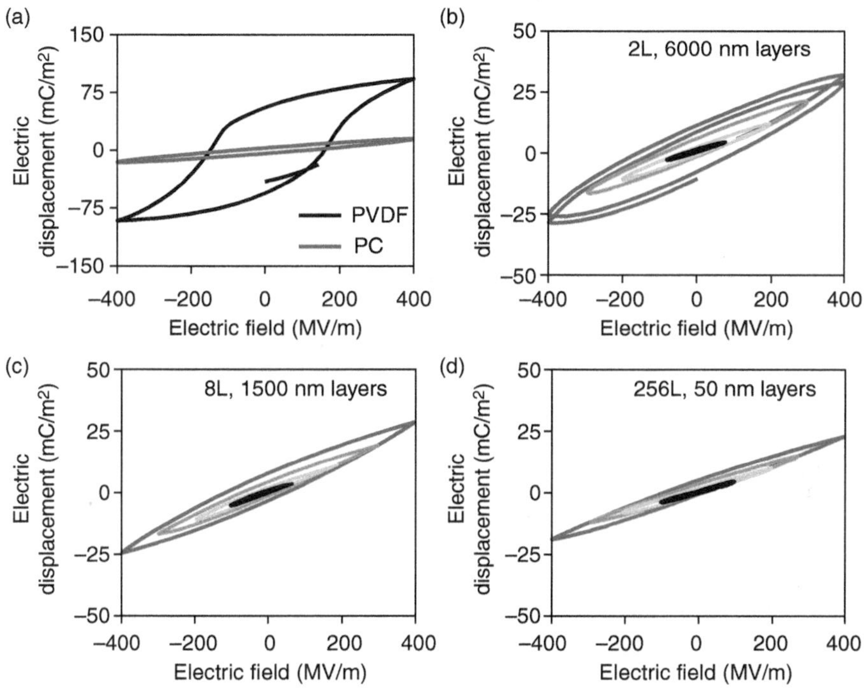

Figure 5.16 D–E hysteresis loops of monolithic controls and 50/50 PC/PVDF layered films with different layer thicknesses. The samples are (a) monolithic PC and PVDF controls, 12 μm thick; (b) 2 layers (12 μm/ 6000 nm); (c) 8 layers (12 μm/1500 nm); (d) 256 layers (12 μm/50 nm). Each loop at a given electric field has two poling cycles. For parts b–d, the increasing size of loops correspond to the poling field being 100, 200, 300, and 400 MV/m, respectively.

the PC layers. When the PVDF layers were significantly thicker than the PC layers, the hysteresis behavior of films resembled PVDF control. As the ferroelectric PVDF control exhibited large dielectric hysteresis due to dipole orientation, the absence of hysteresis in multilayer films may have originated from the suppressed dipole orientation.

Dipole switching in PVDF causes rotation of ferroelectric crystals under applied field resulting in large hysteresis in PVDF controls. The crystal orientation in PVDF layers showed on-edge or weak on-edge crystal orientation with the CH_2CF_2 dipoles perpendicular to the chain axis and parallel to the applied electric field, which should increase the hysteresis [21]. However, the layered films showed reduced hysteresis, which could not be explained as a function of crystal orientation. In addition, the reduced hysteresis could not be attributed to the crystallite size and crystallinity, as these properties were not affected significantly as a function of layer thickness. Further electric field-dependent polarized Fourier transform infrared (FTIR) also did not show any evidence of dipole switching in layered films.

This unique behavior of PVDF layers at high fields was explained by calculating the effective electric field in individual PVDF and PC layers. Using the relative dielectric constants of 3 [9] and 12 [11] for PC and PVDF, respectively, the nominal electric fields for a PC/PVDF serial capacitor model is given as follows:

$$E_{PC} = E_O\left[\phi_{PC}\left(1-\frac{\varepsilon_{r,PC}}{\varepsilon_{r,PVDF}}\right)+\frac{\varepsilon_{r,PC}}{\varepsilon_{r,PVDF}}\right]^{-1} \tag{5.4}$$

$$E_{PVDF} = E_O\left[\phi_{PC}\left(\frac{\varepsilon_{r,PC}}{\varepsilon_{r,PVDF}}-1\right)+1\right]^{-1} \tag{5.5}$$

where ε_o is the external electric field, ϕ_{PC} is the volume fraction of PC, and ϕ_{PVDF} is the volume fraction of PVDF. For example, for an applied external field of 200 MV/m in 50/50 PC/PVDF, the calculated nominal fields in PC and PVDF layers are 320 and 80 MV/m, respectively. The calculated nominal field in PVDF layers was too low to switch dipoles in the layered films. It was also suggested that the coercive field could increase with decreasing PVDF layer thickness making PVDF dipole flipping difficult. It may be possible to induce dipole flipping at significantly higher fields. It was concluded that as the dipole flipping in PVDF did not cause the hysteresis reduction, ionic and interfacial polarization in polymers were considered.

5.2.5.2 Ionic Polarization

Broadband dielectric spectroscopy was used to measure the effect of layering on the ionic polarization as shown in Figure 5.17 [19]. Dielectric spectroscopy measurements were performed in 0.001 Hz to 100 kHz frequency range as a function of number of layers. The films showed two types of dielectric relaxations in 0.001–10 Hz and 10–100 kHz frequency range. The high frequency relaxation peaks in both PVDF control and the multilayered films were associated with the PVDF α-phase relaxation [21]. This relaxation was not affected significantly with increasing number of layers. In contrast, the low frequency relaxation peak, characteristic of ion migration, [22] decreased with the increasing number of layers. The results were compared with the PVDF and PC control to understand the

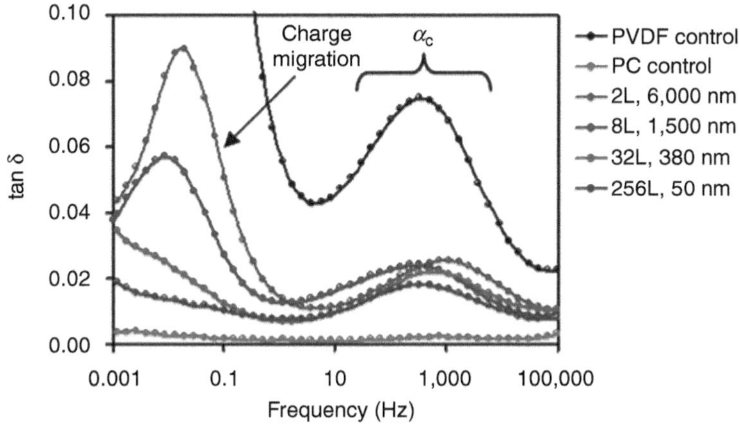

Figure 5.17 Dielectric spectroscopy of 50/50 PC/PVDF layered films at 75°C for different layer thicknesses.

origin of ions in the layered film. The high tan δ values for PVDF control films suggested that the impurity ions were predominantly present in the PVDF layers, while PC control showed relatively flat peak for ion migration. The ions in the PVDF layers may have resulted from surfactant residues during the suspension polymerization-based manufacturing process. Assuming the confinement of the ions to the PVDF layers, the possibility of PC layers and the interfaces as blocking electrodes was suggested due to the higher resistivity of PC (10^{17} Ωcm) than the PVDF (10^{14} Ωcm).

A standard sinusoidal waveform combined with a DC bias voltage ranging from 0 V to 36 V was used to investigate the low frequency behavior of multilayered films. The decreased low frequency peak intensity with increasing voltage supported the charge migration mechanism in the layers rather than the interfacial polarization in the layers [23]. The diffusion coefficients and ion concentration were obtained using Sawada and Coelho diffusion models for PC/P(VDF–HFP) multilayered systems, as shown in Table 5.4. When the layer thicknesses were less than 430 nm, the diffusion coefficient and the ion concentrations were 2×10^{-13} m²/s and 2×10^{21} ions/m³, respectively. In contrast, an order of magnitude higher diffusion coefficient and an order of magnitude lower ion concentration were observed in thicker layers. The estimated values of ion concentrations require further research to understand the correlation between ion concentration and layer thicknesses.

The general ion migration mechanism was visualized as shown in Figure 5.18. When an electric field is applied, the positive and negative charges migrate to interface thereby producing a drift current and ion

Table 5.4 **Diffusion Coefficients and Ion Concentration in PC/P(VDF–HFP) (50/50) Films Obtained by Fitting Experimental Loss Peak Frequency and the Loss Peak Amplitude at 100°C**

No. of Layers	Layer Thicknesses (nm)	D (m²/s)	n (no. of ions/m³)	Loss Peak Frequencies (Hz)	Loss Peak Amplitude
1	14,000	1.6×10^{-12}	5.3×10^{19}	0.013	230
1	7,000	1.2×10^{-12}	1.1×10^{20}	0.04	120
2	7,200	2.6×10^{-12}	7.8×10^{19}	0.08	90
32	430	2.6×10^{-13}	1.9×10^{21}	2	8.3
32	190	1.1×10^{-13}	3.1×10^{21}	5	3
256	50	3.2×10^{-13}	1.0×10^{21}	200	0.5

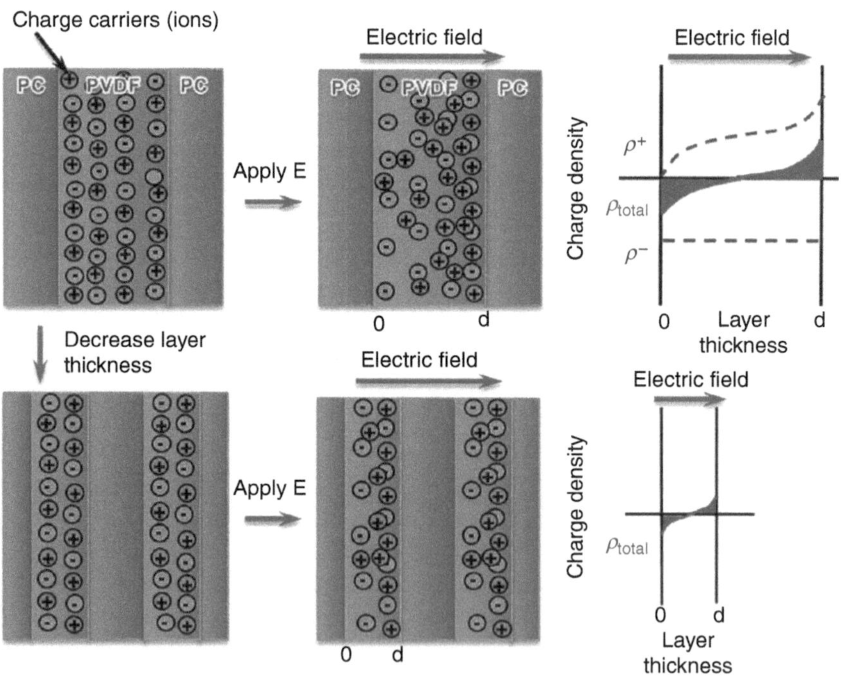

Figure 5.18 Schematic of charge migration in layered films: (top row) thick layers and (bottom row) thin layers. Both positive and negative ions are needed to maintain a zero net charge in the films. For simplicity, it is assumed that only one of the charge carriers moves. After an electric field is applied, there is a gradient in charge concentration. Compared with thick PVDF layers, thin PVDF layers have less charge motion and total charge buildup.

concentration gradient. The charge build up process occurs to balance drift current and diffusion current. In thinner films, charge buildup and amount of ion motion at the interfaces are lower, reducing the hysteresis at high fields and the dielectric loss peak at low fields. Small charge separation in thinner layers also allows faster ion relaxation than the thick layers.

5.2.5.3 Interfacial Polarization

Because multilayered films consisted of two components with high contrast in dielectric constant ($\varepsilon_{r,PC} = 3$ and $\varepsilon_{r,PVDF} = 12$) and the bulk conductivity ($\sigma_{PC} = 10^{-15}$ s/m and $\sigma_{PVDF} = 10^{-12}$ s/m), the Maxwell–Wagner–Sillars (MWS) interfacial polarization built-up at the interfaces under applied electric field [24]. The interfacial polarization is typically caused by the charge carriers such as electrons or holes at the interfaces. As the bulk conductivity of PVDF is higher than the PC controls, the space charges such as ions and electrons should be present at higher concentration in PVDF layers. The estimated ion mobility in PVDF was at least four orders of magnitude smaller than the electron mobility at room temperature. Therefore only electrons were discussed as space charges to explain interfacial polarization [25]. When the electric field is applied, the electrons in PVDF accumulate at the PC/PVDF interfaces with PC as the blocking electrodes. These interfacial charges form effective traps for the electrons from the electrodes.

The phenomenon of electronic conduction in polysulfone (PSF)/PVDF multilayered films was used to investigate interfacial polarization. The 30/70 volume composition of polysulfone (PSF)/PVDF multilayer film was selected due to high interfacial polarization in PVDF layers and relatively easy internal electronic conduction in PSF layers. The PSF/PVDF film samples with varying layer thicknesses are shown in Table 5.5. Effect

Table 5.5 PSF/PVDF (30/70) Multilayered Film Information

No. of Layers	Total Film Thicknesses (μm)	PSF Layer Thickness [nominal/measured] (nm)	PVDF Layer Thickness [nominal/measured] (nm)	PSF/PVDF (v/v)
32	11.6 ± 0.9	218/244 ± 25	508/519 ± 42	29/71
32	9.5 ± 0.5	178/206 ± 24	415/433 ± 27	27/73
32	4.6 ± 0.2	86/83 ± 6	201/241 ± 18	33/67
256	12.0 ± 0.9	28/NA	66/NA	27/73

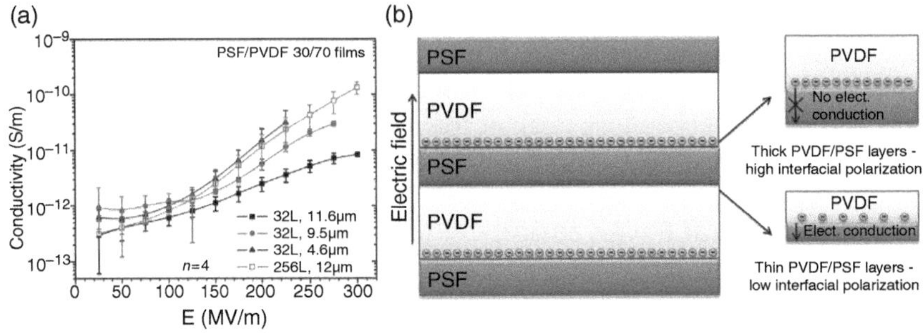

Figure 5.19 (a) Conductivity as a function of electric field (E) for 32 and 256L PSF/PVDF 30/70 films with different total thicknesses. Every data point is an average of four measurements ($n = 4$); and (b) schematic of the MWS interfacial polarization in PVDF layers and subsequent electronic conduction in PSF layers for the PSF/PVDF multilayer films. Here, the space charges are free electrons in PVDF.

of room temperature conductivity as a function of applied electric field on multilayered film samples is shown in Figure 5.19a. Although the conductivity differences were small at low applied fields, noticeable differences with increasing electric field were noticed. For example, a 32-layered film with thicker layers, showed about an order of magnitude lower conductivity at 300 MV/m field, than films with thinner layers. In general, when the PSF layer thickness is greater than 100 nm, the electronic conduction was prevented resulting into higher interfacial polarization and low electrical conductivity. In contrast, when the PSF layer thicknesses are less than 100 nm (PVDF thicknesses were also much thinner), enhanced electronic conduction through the PSF layers reduced the interfacial polarization. The schematic of the suggested MWS interfacial polarization based on PSF/PVDF film system is shown in Figure 5.19b. Such changes in the conductivity properties significantly impacted the dielectric properties such as breakdown strength and dielectric lifetimes as shown in Figure 5.20. The characteristic breakdown strength at the 63.2% failure probability decreased with layer thickness reduction (i.e., reduced polarization). With reducing PSF and PVDF layer thicknesses, the dielectric breakdown strength decreased from 416 MV/m in 11.6 μm films with 250 nm PSF layers to 320 MV/m in 4.6 μm thick films with 80 nm PSF layers. A similar change was observed when the number of layers increased from 32 to 256 while the film thickness stayed constant. The 63.2% Weibull lifetime for 256-layered film was close to 3000 s, which was significantly lower than the 32-layered film Weibull lifetime of approximately 6800 s. Other factors affecting the magnitude of interfacial polarization include the PVDF thickness at higher charge mobility and long poling time, provided PVDF

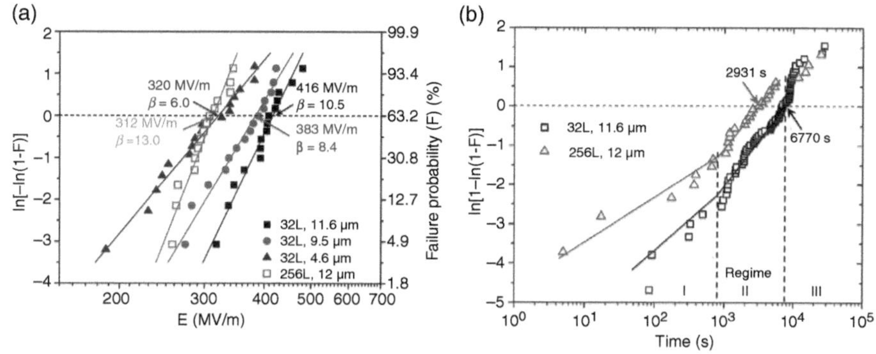

Figure 5.20 (a) Weibull plots of the first breakdown strength distribution for 32- and 256-layered PSF/PVDF 30/70 films with different total film thicknesses; and (b) Weibull plots of the lifetime distribution for the 32- and 256-layered PSF/PVDF 30/70 films.

layers are thin enough. In contrast, when the PVDF layers are thick, space charge mobility is low, and poling times are short, the interfacial polarization will be independent of PVDF layer thickness (above 50 nm thicknesses). Therefore, the relative thicknesses of the two layers are important in optimizing the interfacial polarization for improved properties. In earlier discussed multilayer films with thick PC layers, the electronic conduction in the layers can be prevented to achieve higher interfacial polarization for effectively trapping electrons and reducing conductivity in thicker layers.

5.2.6 Energy Densities in PVDF-Based Multilayered Films

Similar to P(VDF–HFP) copolymer, PVDF homopolymer, with high dielectric constant also demonstrated improved energy density in multilayer structure. The D-E hysteresis loops were integrated to calculate the input, output, and unrecovered energy density [19]. The energy density values of multilayered films were consistently higher in 1.5 μm or thinner layers due to increased breakdown strength. This effect was consistent with observations in PC/P(VDF–HFP) multilayer films. For PC/PVDF system with 380 nm layers, the observed charge energy density of 11 J/cm^3 was higher than PC and PVDF controls as well as other layered films. Increased energy densities correlated with enhanced breakdown strength as evident from the maximum electric field in Figure 5.21. D-E hysteresis loops were also used to calculate the unrecovered energy as shown in Figure 5.21b. PVDF control films showed highest unrecovered energy, while PC control showed significantly lower unrecovered energy. In addition, the layered films showed reduced unrecovered energy, which decreased

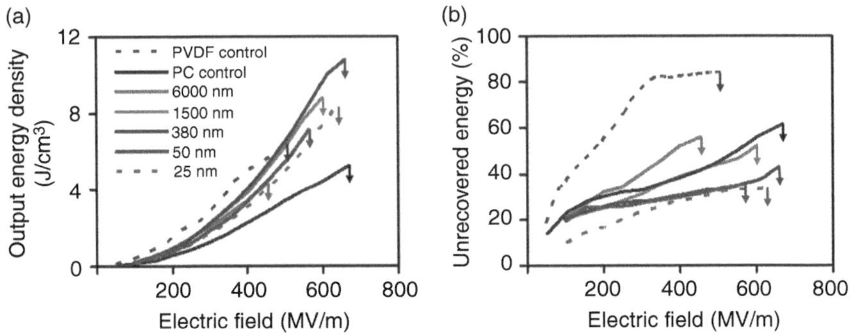

Figure 5.21 (a) Output energy density and (b) percent unrecovered energy of 50/50 PC/PVDF films as a function of electric field.

with decreasing layer thickness. For layer thickness below 1500 nm, the amount of unrecovered energy was less than PC control.

5.2.7 Interface Modification in Multilayered Films

The breakdown mechanism and various polarizations revealed the significance of the interfaces in improving the dielectric properties in layer composites. The layer interfaces acted as barriers to breakdown propagation improving the breakdown strength and the dielectric lifetimes. Relative layer thicknesses and interfaces played a crucial role in controlling polarization effects.

To further evaluate the effect of interfaces on dielectric properties, an earlier discussed approach of tie layer materials was used to vary adhesion, interphases, and/or interdiffusion properties [26]. Details of the film production process utilizing the tie layers are discussed in Chapter 2. As discussed in Chapter 3, tie layers can be used to tune the mechanical behavior of the films. Multilayered films of PC/P(VDF–HFP) with different tie layers and layer thicknesses were produced in $[ATBTA]_n$ layer configuration, as shown in Table 5.6. The tie layer concentration varied from 2 vol% to 33 vol%, while the ratio of PC and P(VDF–HFP) materials was kept constant as 1:1. The total number of PC and P(VDF–HFP) layers were maintained at 33 in the film, while the addition of tie layers increased the total number of layers to 65. Three tie layer materials used were – PMMA, SAN30, and poly(ethylene terephthalate-*co*-1,4-cycohexanedimethylene terephthalate) (PETG). These materials were selected for their unique interaction with PC and P(VDF–HFP) layers. The interactions of the three tie layer materials with PC and P(VDF–HFP) layers are shown in Table 5.7.

Table 5.6 PC/Tie/P(VDF–HFP) Multilayer Film Compositions

Multilayer Film Compositions: PC/Tie/ P(VDF–HFP) (vol%)	No. of Layers	Tie Layer Thicknesses (nm)	PC or PVDF Layer Thicknesses (nm)
50/0/50	33	–	312
49/2/49	65	6	306
48/4/48	65	13	300
46/8/46	65	25	288
40/20/40	65	63	250
33/33/33	65	103	206

Table 5.7 Tie Layer Components and Their Interactions with PC and P(VDF–HFP) Layers

Multilayer Film Components	Tie Layers		
	PMMA	SAN30	PETG
PC	Strong interphase	Immiscible	Miscible
P(VDF–HFP)	Miscible	Possible interaction	Immiscible

5.2.7.1 Effect of PMMA Tie Layers

A significant impact of PMMA tie layers on the breakdown strength of the layered films was observed. The effect of PMMA tie layer thicknesses on the dielectric breakdown strength is shown in Figure 5.22a. When PMMA tie layer thickness was 25 nm, a 25% improvement in the breakdown strength (from 700 kV/mm in absence of PMMA layers to 880 kV/mm in films with PMMA layers) was achieved using a needle–plane measurement. Further, an increase in the nominal PMMA layer thickness decreased the breakdown strength slowly. It is obvious from Equation (5.1) that the improved breakdown strength can increase the maximum discharge energy density. The energy density values calculated from the D-E polarization hysteresis measurements are shown in Figure 5.22b. The maximum energy density for 33 layered 50/50 PC/P(VDF–HFP) film and 46/8/46 PC/PMMA/P(VDF–HFP) was 5.7 and 8.4 J/cm^3, respectively. In addition, the PMMA tie layers also reduced the loss (unrecovered) energy density. For example, the loss (unrecovered) energy density of 4.8 J/cm^3 in PC/P(VDF–HFP) films without the

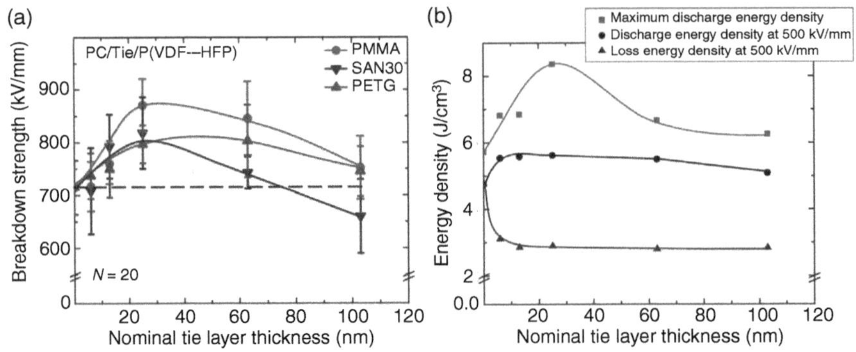

Figure 5.22 (a) Dielectric breakdown strength as a function of nominal tie layer thickness for 65-layered PC/PMMA/P(VDF–HFP) films. The 33-layered PC/P(VDF–HFP) film was also included and taken as the film with 0-nm tie layer thickness (indicated as the dashed line). (b) Maximum discharge energy density and hysteresis property values at 500 kV/mm as a function of PMMA tie layer thickness for the 65-layered PC/PMMA/P(VDF–HFP) multilayer films.

PMMA tie layers reduced by 40% to 2.8 J/cm^3. These reduced losses were associated with the suppressed ion migration as confirmed by low-field dielectric spectroscopy as shown in Figure 5.23. As the room temperature testing did not show any ion migration peaks at frequencies 0.01 Hz and above for films containing PMMA, a 100°C test temperature was used to shift the ion migration peak to higher frequency. At the testing temperatures, the α-phase relaxation peak drifted to higher frequencies beyond 10^5 Hz, while the ion migration peaks were visible in the testing frequency range. The ion migration peak for P(VDF–HFP) in the multilayered films without the tie layers was visible at 30 Hz. When PMMA was present in the composite system, this peak shifted to lower frequencies indicative of suppressed ion migration.

It was suggested that the localized mixing of PMMA layers with P(VDF–HFP) created interphases, which acted as traps for ions reducing the ion mobility. Alternative explanation proposed restricted ion movement across the interface due to the interphase region formed by PMMA intermixing with the amorphous P(VDF–HFP) chains [27]. The evidence of the interphase formation is demonstrated by AFM imaging in Figure 5.24a. AFM images of the films showed sharp boundaries between PC and P(VDF–HFP) layers in 33-layered 50/50 film. The absence of distinct PMMA layers and a diffuse interphase boundary at 8% concentration indicated PMMA interdiffusion with PC and P(VDF–HFP) layers. Films with higher concentrations of PMMA showed a distinct PMMA layer with the thicknesses significantly less than target thicknesses and

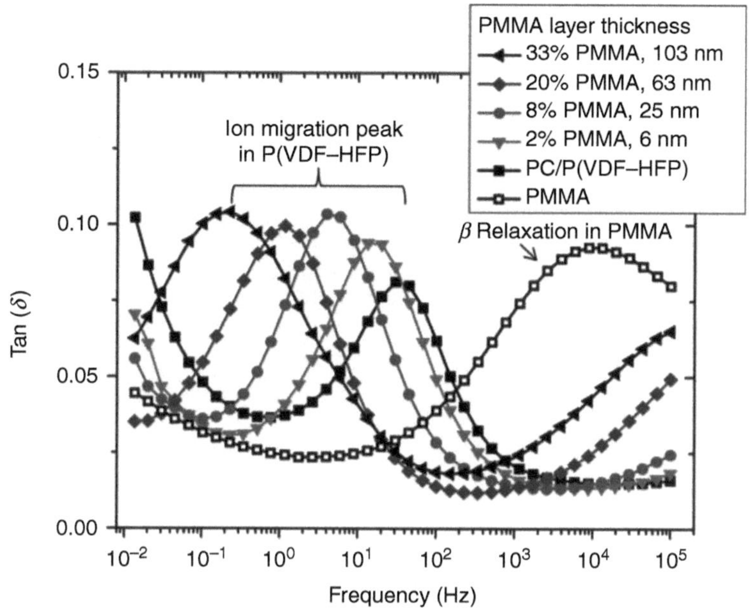

Figure 5.23 Dielectric loss tangent of 65-layered PC/PMMA/P(VDF–HFP) films in addition to 33-layered PC/P(VDF–HFP) and PMMA controls as a function of frequency at 100°C. All films had equal amount of PC and P(VDF–HFP), and film thicknesses were maintained at 10 μm.

adjacent interdiffusion regions. Such formations of interphase materials are discussed in Chapter 3. Due to the differences in relative solubility parameters of the polymer pairs, thicker interphase of PMMA/P(VDF–HFP) and thinner interphase of PC/PMMA were suggested in this film. A schematic representation of the PC/PMMA/P(VDF–HFP) films as a function of PMMA layer thickness is shown in Figure 5.24b.

5.2.7.2 Effect of PETG and SAN30 Tie Layers

In contrast to PMMA adhesion with PC and P(VDF–HFP), PETG, and SAN30 showed weaker interactions with at least one of the two components in the multilayered film. Films with SAN30 tie layers showed sharper layer boundaries as compared to PETG layers. Stronger interaction of PETG with PC created interphase material at PETG/PC interface, while weaker interaction with P(VDF–HFP) produced sharp interfaces. In contrast, SAN25 tie layers showed discrete layers without significant interphase formation with both PC and P(VDF–HFP). It was confirmed that the measured PETG layer thicknesses were less than the target values

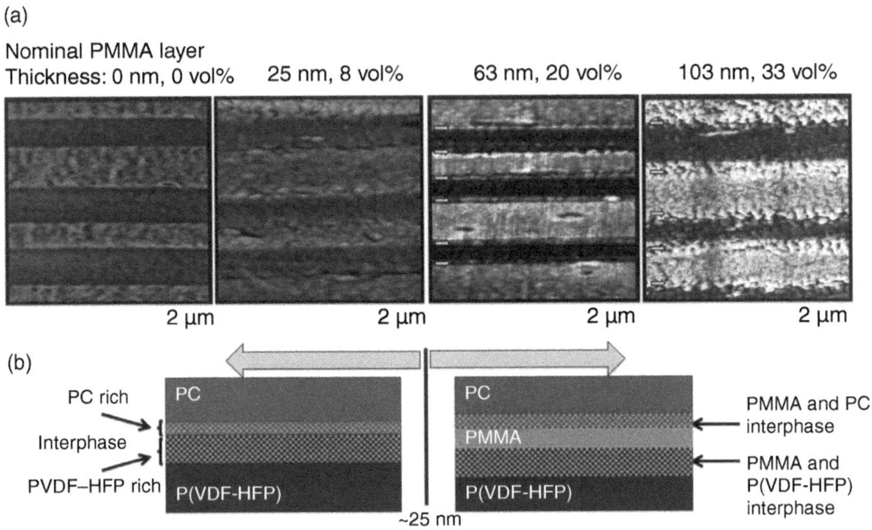

Figure 5.24 (a) AFM images of selected PC/PMMA/P(VDF–HFP) 65-layered samples. Example PMMA thicknesses are 25, 63, and 103 nm. PC/P(VDF–HFP) control is included for comparison. (b) Schematic representation of the layered structure in the PC/PMMA/P(VDF–HFP) 65-layered samples as a function of tie layer thickness.

indicative of interphase formation, while SAN30-layered thicknesses were comparable with the target values. As expected, with increasing layer thickness, the interphase fraction reduced significantly and the tie layer materials were clearly visible in the AFM images, as shown in Figure 5.25. The contrasting profile of the AFM images produced a relative gray value profile across the images. The transition of the gray values showed broad transition in PMMA containing film confirming stronger interaction of PMMA with PC and P(PVDF–HFP). Sharper transitions for PETG and SAN30 materials with PC or P(VDF–HFP) indicated weaker interactions between these polymers. Therefore, the improvement in the breakdown strength in PETG and SAN30 containing films was not as strong as multilayer films with PMMA.

The discharge energy densities calculated from D-E hysteresis loops are shown in Table 5.8. The maximum discharge energy density showed some improvement in PETG containing films, while properties deteriorated slightly when SAN30 was present as tie layers. Similar to PMMA tie layers, the loss energy density and hysteresis losses decreased with PETG and SAN30 tie layers. Both PETG and SAN30 also contributed to suppressed ion migration. As shown in Figure 5.26, the intensity of an ion migration peak in the films with PETG and SAN30 was higher than PMMA-based

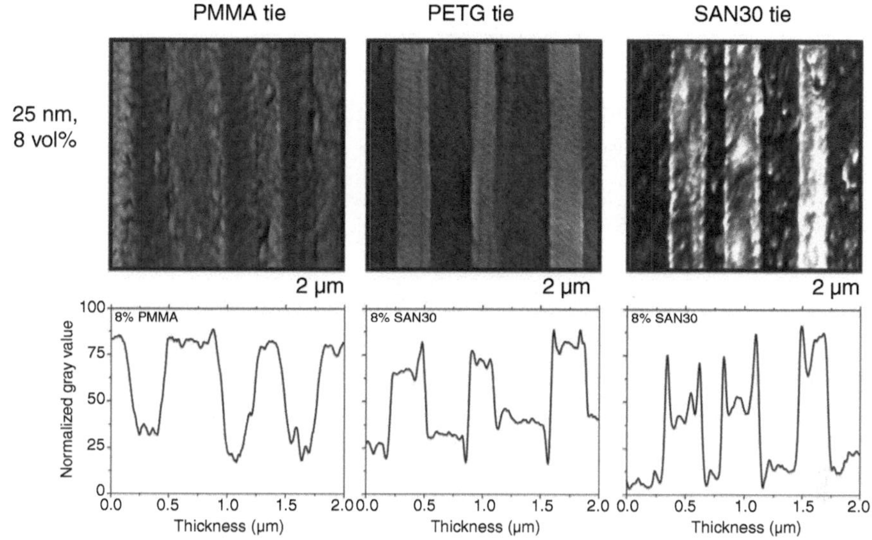

Figure 5.25 AFM images (upper) and their gray-value profiles (lower) of selected PC/tie/P(VDF–HFP) 65-layered samples with 25-nm tie layer thickness. Example tie layers are PMMA (left column), PETG (middle column), and SAN30 (right column). Note: the images are turned by 90°.

films. Also the ion migration peak showed highest shift with PMMA tie layers, as compared to PETG and SAN30 tie layers. When the films were tested at 100°C, the ion migration peak observed at 30 Hz frequency for PC/P(VDF–HFP) films changed to 4 Hz frequency in films with PMMA tie layers, while PETG and SAN30 tie layers changed the peak frequency only slightly to 40 and 25 Hz, respectively.

Table 5.8 Effect of Tie Layer Materials on Discharge Energy and Loss Energy Density Values in PC/Tie/P(VDF–HFP) (46/8/46) Multilayer Films with a Nominal Tie Layer Thickness of 25 nm

Tie Layer Materials	Maximum Discharge Energy Density (J/cm^3)	Discharge Energy Density (J/cm^3)	Loss Energy Density at 500 kV/mm (J/cm^3)
No tie	5.72	4.75	4.79
PMMA	8.36	5.62	2.9
PETG	6.11	5.07	3.79
SAN30	4.86	4.86	3.85

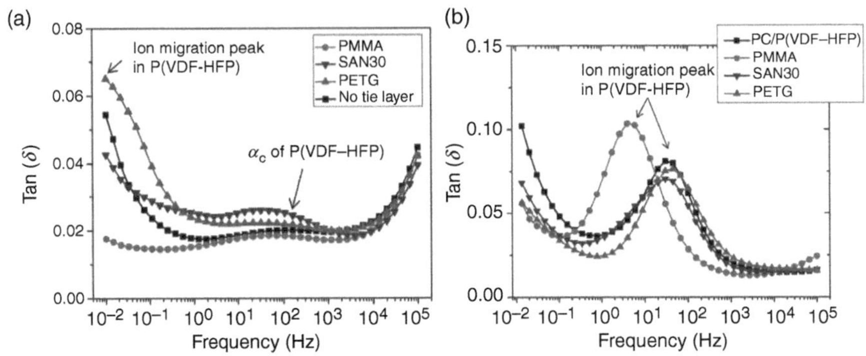

Figure 5.26 (a) Dielectric loss tangent of various 46/8/46 PC/tie/P(VDF–HFP) 65-layered films. The 50/50 PC/P(VDF–HFP) 33-layered was also included for comparison. All tests are done at 20°C; and (b) dielectric loss tangent as a function of frequency for various 46/8/46 PC/tie/P(VDF–HFP) 65-layered films in addition to the 50/50 PC/P(VDF–HFP) 33-layered film. All measurements were conducted at 100°C.

Tie layer materials also impacted the dielectric constant of the PC/P(VDF–HFP) films and the effective dielectric constant of P(VDFvHFP) layers. The P(VDF–HFP)/tie interactions were important than PC/tie in determining P(VDF–HFP) effective dielectric constant, which was calculated using a series model shown in Equation (5.6).

$$\frac{1}{\varepsilon_{\text{calc}}} = \frac{\phi_{\text{tie}}}{\varepsilon_{\text{tie}}} + \frac{\phi_{\text{PC}}}{\varepsilon_{\text{PC}}} + \frac{\phi_{\text{PVDF-HFP}}}{\varepsilon_{\text{PVDF-HFP}}} \tag{5.6}$$

where $\varepsilon_{\text{calc}}$, ε_{tie}, ε_{PC}, and $\varepsilon_{\text{P(VDF–HFP)}}$ are the dielectric constants of the multilayer films, tie layer materials, PC control and P(VDF–HFP) control, respectively. ϕ_{tie}, ϕ_{PC}, ϕ_{PVDF} are the fractions of the tie, PC, and P(VDF–HFP) polymers in the film. The measured dielectric constant, the calculated dielectric constant and the effective dielectric constant for 46/8/46 PC/tie/P(VDF–HFP) films are shown in Table 5.9. PMMA tie layer system showed the highest effective dielectric constant for P(VDF–HFP). The possibility of PMMA as a lubricating agent to loosen up the P(VDF–HFP) dipoles or formation of β-phase of P(VDF–HFP) with higher dielectric constant was suggested [18,28]. The SAN30 tie layers with intermediate improvement in the dielectric constant were attributed to its weaker interaction with P(VDF–HFP). Unlike PMMA and SAN30, the series model showed that the calculated dielectric constant for PETG containing films was similar to measured values due to its weaker interactions with P(VDF–HFP).

Table 5.9 **Measured and Modeled Dielectric Constants of PC/Tie/P(VDF–HFP) 46/8/46 Multilayer Films at 20°C, and the Effective Dielectric Constant of P(VDF–HFP) Layers Calculated From the Series Model**

Film Systems	Compositions	Measured Dielectric Constants	Calculated Dielectric Constant	Effective Dielectric Constant of P(VDF–HFP)
PC/P(VDF–HFP)	50/50	4.46	4.49	10.8
PC/PMMA/ P(VDF–HFP)	46/8/46	4.98	4.42	29.6
PC/SAN30/ P(VDF–HFP)	46/8/46	4.56	4.3	16.42
PC/PETG/ P(VDF–HFP)	46/8/46	4.49	4.38	12.96

5.2.8 Biaxially Oriented Multilayered Films

Biaxial stretching approach is conventionally used to produce commercial PP or PET films for dielectric applications. As discussed in Chapter 3, BOPP films modified by incorporating PEO or PCL nanolayers improved the barrier properties significantly without sacrificing the mechanical or optical properties [29]. This approach of nanolayer incorporation in the commercial polymers was extended to PET dielectric films. In the first step, the synergistic effects of high dielectric constant PVDF-based polymers with high breakdown strength polymers were extended to incorporate PVDF-copolymer in PET films. Subsequent biaxial stretching of layered films produced thin PET films with PVDF-copolymer nanolayers.

A PVDF copolymer, poly(vinylidene fluoride-*co*-tetrafluoroethylene) (PVDF–TFE) was coextruded with PET to produce a multilayer film structure with 32 layers. The film compositions and corresponding layer thicknesses produced have been shown in Table 5.10. The effect of biaxial orientation on the crystalline morphologies and the corresponding dielectric properties was investigated. Biaxial stretching of 200 μm films was carried out at 105°C to 4.5 × 4.5 draw ratios at a stretch rate of 100%/s to achieve final film thickness close to 10 μm. The final layer thicknesses varied between 60 nm and 560 nm after stretching the multilayered films. Changes in the confined PET and PVDF–TFE structures have been shown in Figure 5.27. PVDF–TFE layers contained randomly oriented crystals,

Table 5.10 PET/P(VDF–TFE) 32-Layered Multilayer Films Under Investigation

Compositions PET/P(VDF–TFE) (v/v)	Layer Thicknesses (nm)
0/100	(P(VDF–TFE)
10/90	60/560
20/80	125/500
30/70	190/440
40/60	250/375
50/50	310/310
60/40	375/250
70/30	440/190
80/20	500/125
90/10	560/60
100/0	PET control

while PET layers were amorphous in extruded films. Biaxial stretching at 105°C produced on-edge morphology in PVDF–TFE layers, whereas PET layers formed oriented fibrils. Furthermore, two thermal annealing approaches were also used to create different morphological features in the layer components. The first thermal annealing approach involved annealing of films at 140°C, followed by quenching to room temperature, which formed oriented and crystallized PET fibrils. In the second approach, a subsequent step to selectively melt the PVDF–TFE layers at 140°C (above $T_{m,\text{PVDF–TFE}}$ = 125°C) after biaxial stretching and isothermal crystallization at 120°C ($T_{c,\text{PVDF\_TFE}}$ = 110°C) for 3 h was performed to achieve in-plane lamellar orientation in PVDF–TFE layers. The annealing step also caused the crystallization of oriented PET fibrils. The four sets of multilayered films with different thermal history and crystal orientations were analyzed to optimize the dielectric performance. The effect of thermal annealing on the structural changes in multilayered film has been summarized in Figure 5.27c.

Breakdown strength of the films with different processing conditions, as a function of PET (or PVDF–TFE) composition, measured by needle–plane set-up, is shown in Figure 5.22. Similar to earlier discussed films, the PET/PVDF–TFE coextruded films showed improved properties as compared PET and PVDF–TFE control films and the weighted average model. The maximum improvement was observed in 70/30 PET/PVDF–TFE films. Further improvement in the range of 25–40% was achieved by biaxial stretching of the thick films. Although the subsequent thermal

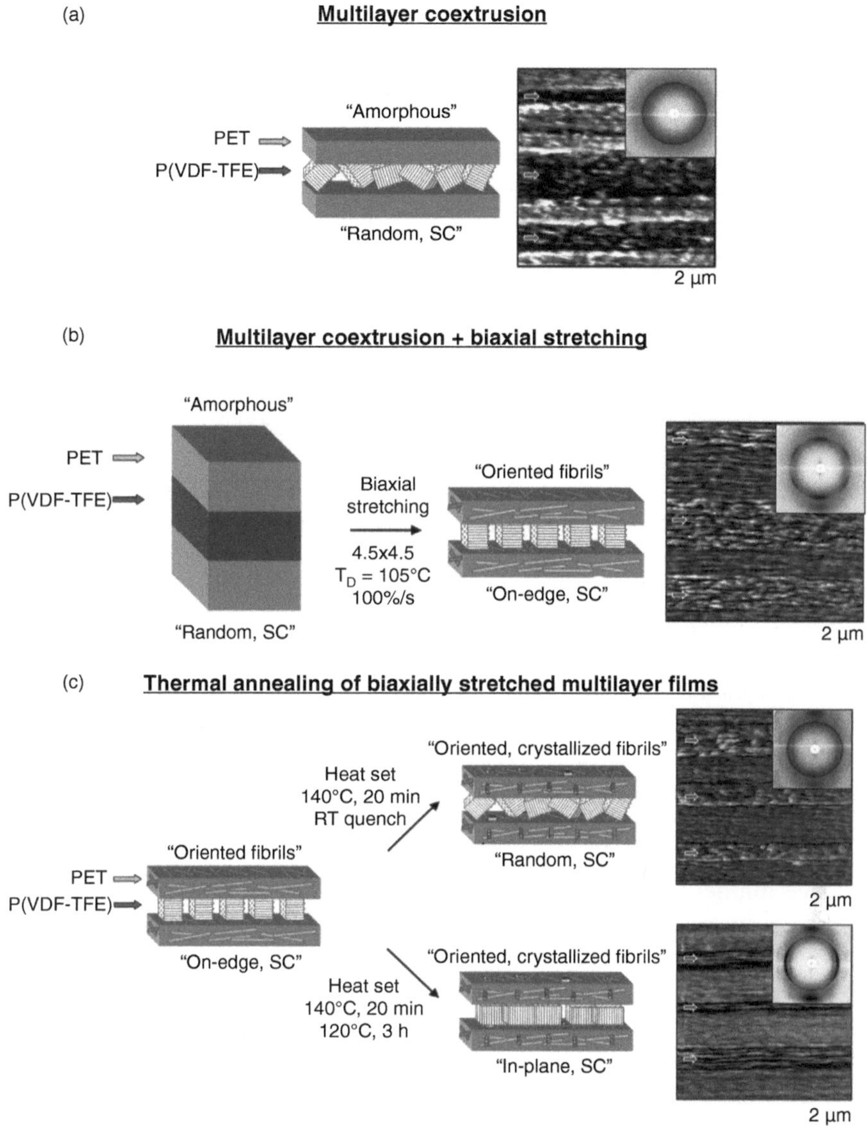

Figure 5.27 Structures obtained in the PET/P(VDF–TFE) multilayer system. Through (a) multilayer coextrusion and (b) biaxial stretching of thick PET/P(VDF–TFE) multilayer films at high draw ratios. The wide-angle X-ray scattering images were taken through the extrusion direction of the multilayer films. The arrows in the atomic force microscopy images indicate the P(VDF–TFE) layers (c) Thermal annealing approaches on the biaxially stretched PET/P(VDF–TFE) multilayer films to produce different structures in the PET and P(VDF–TFE) layers. Approaches are heat isothermally at 140°C for 20 min, then quench to room temperature and heat isothermally at 140°C for 20 min, heat isothermally at 120°C for 3 h, and finally cool slowly to room temperature. The wide-angle X-ray scattering images were taken through the extrusion direction of the multilayer films. The arrows in the atomic force microscopy images indicate the P(VDF–TFE) layers.

annealing processes showed improvement in breakdown strength over the coextruded films, biaxially stretched (without any thermal treatment) films still showed the maximum improvement in breakdown properties. Improved energy density properties associated with enhanced breakdown strength have been shown in Figure 5.22. Biaxially oriented films showed energy density values as high as 16 J/cm^3, which represents a twofold increase over biaxially stretched PET control films. Interestingly, thermally annealed films did not show any improvement in the energy density.

Similar to earlier results, damage zones in the control films showed single breakdown holes, while layered films revealed distinct treeing patterns in a divergent electric field. A minimum of 500 kV/mm-applied field was required to detect the treeing patterns in 10 μm thick films. The results were consistent with earlier discussed PC/P(VDF–HFP) (50/50) film system, which showed treeing above 500 kV/mm-applied field in 12 μm films and close to 600 kV/mm for 9 μm film. The effect of tree diameter as a function of electric field has been shown in Figure 5.29. The average tree diameter increased linearly with increasing field in extruded film and leveled off at 750 kV/mm. In biaxially stretched films, the average tree diameters showed a linear increase up to 1100 kV/mm electric field. In contrast, the thermal treatment of the biaxially stretched films showed negative impact on the tree diameters. Biaxially stretched heat set films showed increasing tree diameters up at 950 kV/mm and subsequent leveling off with further increase in applied field. Irregular tree diameters were observed in the biaxially stretched 120°C recrystallized samples. The changes in the tree diameter and breakdown properties of these films were attributed to crystal morphology in PVDF–TFE layers. In biaxially stretched 120°C crystallized films, the in-plane crystals were present perpendicular to the applied electric field, which allowed charge build-up at crystalline/amorphous interface, within the PVDF–TFE layer and decreased the charge build-up at the PET/PVDF–TFE interface, possibly reducing the treeing phenomena during breakdown events. In contrast, the confined morphology of the PVDF–TFE layers in the biaxially stretched films produced on-edge lamellae, which significantly impacted the charge build-up at the interphase and enhanced breakdown properties as reflected by increased treeing diameter. The treeing mechanism damaged only the surface and a few top layers in the film, which was similar to breakdown mechanism of PC/PVDF films shown in Figures 5.4 and 5.5.

The charge build-up at the interfaces is dependent on the dielectric constant and/or conductivity of the component polymers (Figures 5.28 and 5.29). To understand the improved breakdown properties, the dielectric constants of multilayered films (50/50, PET/PVDF–TFE) was measured and compared with a series capacitor model [9]. The measured dielectric

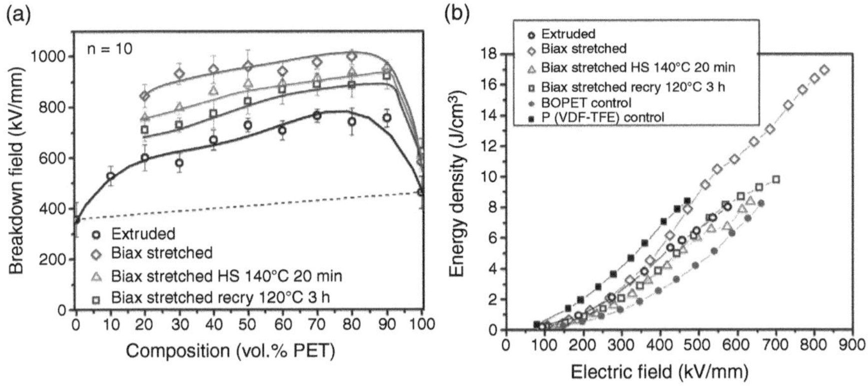

Figure 5.28 (a) Pulsed voltage, needle/plane electrodes, and breakdown field of various 10 mm PET/P(VDF–TFE) 32-layered films as a function of PET composition, (b) energy density of various 10 μm PET/P(VDF–TFE) 32-layered films as a function of composition.

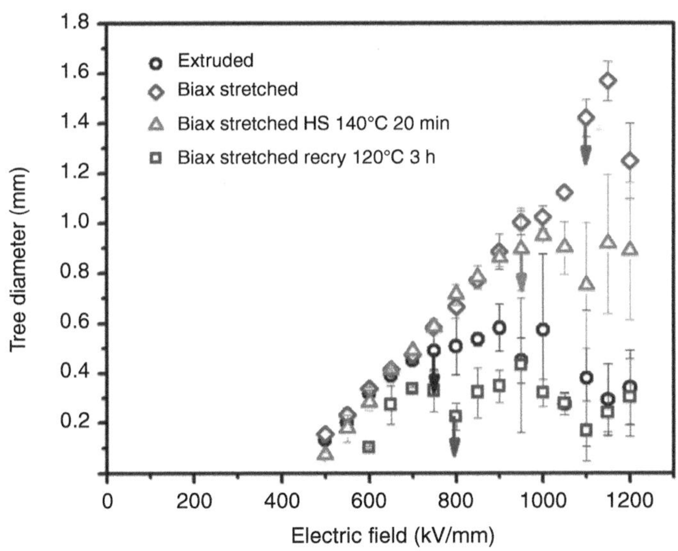

Figure 5.29 Measured tree diameter as a function of electric field for the various PET/P(VDF–TFE) 50/50 32-layered films. The arrows indicate the field at which breakdown occurred.

constant of coextruded multilayer film was 5.2 at 1 Hz, which increased to 5.9 in biaxially stretched films and decreased to 4.9 in thermally treated biaxially stretched films. The increased dielectric constant in biaxially stretched films is consistent with the increased energy density. The impact of the confined morphology on the effective dielectric constant of PVDF–TFE layers was calculated using a series capacitor model for the relative dielectric constant in multilayered composite shown in Equation (5.7).

$$\frac{1}{\varepsilon_{\mathrm{r}}} = \frac{\phi_{\mathrm{PET}}}{\varepsilon_{\mathrm{PET}}} + \frac{\phi_{\mathrm{PVDF\text{-}TFE}}}{\varepsilon_{\mathrm{PVDF\text{-}TFE}}} \tag{5.7}$$

where $\phi_{\mathrm{PVDF{-}TFE}}$ and ϕ_{PET} are the volume fractions of PVDF–TFE and PET, respectively. $\varepsilon_{\mathrm{r,PVDF{-}TFE}}$ and $\varepsilon_{\mathrm{r,PET}}$ are the dielectric constants of PVDF–TFE and PET, respectively. The dielectric constant of PET was assumed to be a constant and same for biaxially stretched PET control. Calculated effective dielectric constant of PVDF–TFE layers in 50/50 films with different thermal histories showed dramatic differences as depicted in Table 5.11. It was as high as 19 in biaxially stretched films and significantly lower (8.5–11.0) in thermally treated multilayered films. The improved dielectric constant was attributed to on-edge crystals in the biaxially oriented films, where the chain axes were parallel to the layers and perpendicular to the applied field. As the dipoles in the PVDF–TFE layers are oriented perpendicular the chain axis, they were present in-plane to the applied electric field. The dipole relaxation under the applied electric field is, therefore, easier to achieve in biaxially stretched films with in-plane crystals. In contrast, dipole relaxation in thermally reset films with in-plane crystals with dipoles perpendicular to the applied field was difficult resulting into reduced effective dielectric constant. Reduced dielectric contrast between PC and PVDF–TFE can also lead to a decreased charge build-up across the interface resulting into reduced breakdown strength and smaller diameter trees.

5.2.8.1 Metal-Filled Conductive Multilayers

Conductive polymers offer unique opportunities in many applications such as portable electronic devices, transducers, conductive adhesives, coatings, tapes, and electromagnetic shielding [30]. Conventional insulating polymers can be filled with metallic fillers to achieve conductivity above certain critical concentrations, which is dependent on the size and shape of the filler. Many approaches to separate the conductive network from the polymer matrix have been used to reduce the required volume concentration. This section discusses multilayer film coextrusion approach to separate conductive layers from nonconductive alternating layers and produce structures with anisotropic conductivity.

Table 5.11 Relative Dielectric Constants of 50/50 PET/P(VDF–TFE) 32-Layered Films and the Respective Calculated Effective Dielectric Constant of the P(VDF–TFE) Layer Using a Series Capacitor Model

Sample: 50/50 PET/P(VDF–TFE) 32 Layers	Overall Film Dielectric Constant at 1 Hz, 25°C	Effective Dielectric Constant of P(VDF–TFE) Layer at 1 Hz, 25°C	Overall Film Dielectric Constant at 100 Hz, 25°C	Effective Dielectric Constant of P(VDF–TFE) Layer at 100 Hz, 25°C
Extruded	5.2	9.9	5	8.7
Biaxially stretched	5.9	19	5.8	17
Biaxially stretched, heat set at 140°C, 20 min	5.3	11	5.2	10
Biaxially stretched, recrystallization at 120°C, 3 h	4.9	8.5	4.9	8

Multilayered composites of PP filled with copper or nickel were fabricated using coextrusion processing [31]. Silver coated copper flakes with an average thickness of 3 μm and aspect ratio of 15, and nickel flakes with an average thickness of 1 μm and aspect ratio of 10 were blended with PP. The concentrations varied from 2.5% to 20% (v/v) for copper and from 5% to 25% (v/v) for nickel. Two types of structures fabricated using coextrusion processing were filled–filled and unfilled–filled composites. In filled–filled (F–F) system, both alternating layers contained metal filled PP to produce 128-layered composites, while the unfilled–filled (U–F) composites consisted of alternating layers of metal filled PP and unfilled PP to produce 8-, 16-, 64-, and 128-layered composites with 10% copper flakes. The U–F samples containing 15% nickel were also produced with 64, 128, and 256 layers. The composite thickness in all samples was 2 mm. The multilayered composites for F–F and U–F composites showed highly oriented metal fillers in direction of the layers. As the average layer thicknesses of 83 and 15 μm, in 16 and 64 layered U–F composites,

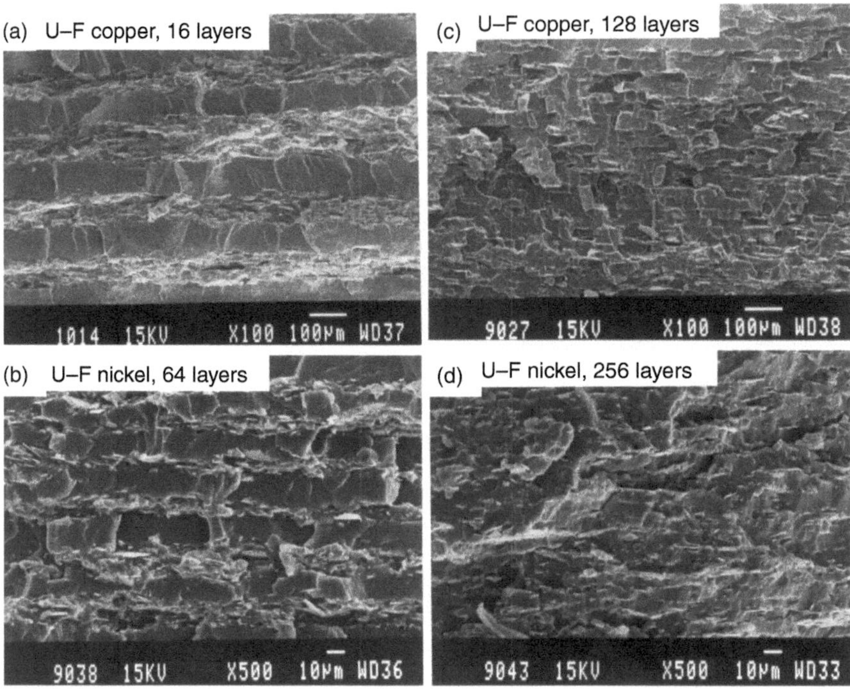

Figure 5.30 Scanning electron micrographs of U–F microlayers with thick layers. (a) Alternating 10% (v/v) copper-filled polypropylene and unfilled polypropylene with 16 layers. (b) Alternating 15% (v/v) nickel-filled polypropylene and unfilled polypropylene with 64 layers. (c) Alternating 10% (v/v) copper-filled polypropylene and unfilled polypropylene with 128 layers. (d) Alternating 15% (v/v) nickel-filled polypropylene and unfilled polypropylene with 256 layers.

respectively, were significantly higher than the copper filler thickness of 3 μm, discernible filled and unfilled layers were observed in the composites as shown in Figure 5.30. However, the layers were indistinguishable as the layer thicknesses became comparable with the filler thickness. For comparison, the compression-molded samples of filled PP prepared from melt blends showed oriented copper and nickel flakes in the direction of molded composite.

The volume resistivity of these composites was measured by applying voltage through the conductive tape and measuring electric current through the sample. From the slope of the linear voltage–current relationship, the resistivity was calculated. The specific resistivity was calculated as

$$\rho = RA / L, \tag{5.8}$$

where A and L are the contact area and the sample thickness, respectively.

In the compression-molded samples, the decreased resistivity of copper and nickel filled composites from 10^{12} Ωcm to 10^2–10^3 Ωcm in unfilled composites, was indicative of electrical conductivity in the composites. The critical volume fractions (percolation threshold) of 4% (v/v) of copper and 8% (v/v) of nickel were required to achieve the electrical conductivity. Below these concentrations, the compression-molded composites were nonconductive. This percolation threshold was significantly lower than 20–40% (v/v) critical volume fraction required for spherical shape particles [32]. In contrast, the higher specific surface area of the copper and nickel platelet particles increased the particle–particle contact probability, resulting in a lower percolation threshold in these systems.

The resitivities of these composites were measured in in-plane and cross-plane directions to investigate the effect of layering on anisotropy properties. In a compressed molded sample, the in-plane resistivity for 15% nickel filled blend composite was only 4–5 times higher than the cross-plane resistivity. The F–F multilayer composites exhibited an order of magnitude anisotropy in the electrical conductivity properties, as shown in Table 5.12. The reduced in-plane resistivity was attributed to filler orientation to the layer direction. The cross-plane resistivity in nickel and copper filled F–F composites was similar to comparable compression molded samples, but in-plane resistivity decreased by an order of magnitude.

Dramatic differences in the resistivity anisotropy were observed in U–F composites. For example, 7.5% nickel filled U–F composites with 64 layers showed in-plane resistivity comparable with the compression molded samples and cross-plane resistivity was similar to unfilled polypropylene. Interestingly, when the number of layers increased to 256, the in-plane conductivity was absent. Similarly, U–F composites with 5% copper showed in-plane conductivity in 8- and 16-layered composites, but became nonconductive in 64- and 128-layered systems. It is interesting to note that thicker filled layers were required for conductive behavior of multilayer composites, while thinner layers produced nonconductive composites. Layer thickness estimations suggested that the composite conductivity was observed when the layer thicknesses comparable to filler particles. Thicker filled layers displayed evidence of a three-dimensional (3D) particle arrangement within the layers, while the thinner layer showed a two-dimensional (2D) layout of metal flakes to demonstrate nonconductive properties.

A schematic describing the development of conductive network in the two sets of composites is shown in Figure 5.31a,b. The conductive network in thick layers resulted from the geometrical contact either from side-to-side contacts or top-to-bottom contacts within filled layers. The

Table 5.12 Resistivities of Metal-Filled Microlayers and Blend Controls

Fillers	Systems	Filler Content% (v/v)	No. of Layers	Layer Thicknesses (μm)	*K* (Layer to Particle Thickness Ratio)	Volume Resistivities (Ωcm) × 10^2	Directions
Ni	Blend	15	–	–	–	27 ± 12	Cross-plane
Ni	Blend	15	–	–	–	6 ± 4	In-plane
Ni	F–F	15	128	7	7	57 ± 11	Cross-plane
Ni	F–F	15	128	7	7	0.43 ± 0.11	In-plane
Ni	U–F	7.5	64	14	14	120 ± 28	In-plane
Ni	U–F	7.5	256	4	4	Nonconductive	In-plane
Cu	Blend	10	–	–	–	2 ± 1	Cross-plane
Cu	F–F	10	128	10	3	0.29v0.05	In-plane
Cu	F–F	10	128	10	3	2.5 ± 1.2	Cross-plane
Cu	U–F	5	8	171	57	0.24 ± 0.01	In-plane
Cu	U–F	5	16	86	29	0.91 ± 0.14	In-plane
Cu	U–F	5	64	21	7	Nonconductive	In-plane
Cu	U–F	5	128	12	4	Nonconductive	In-plane

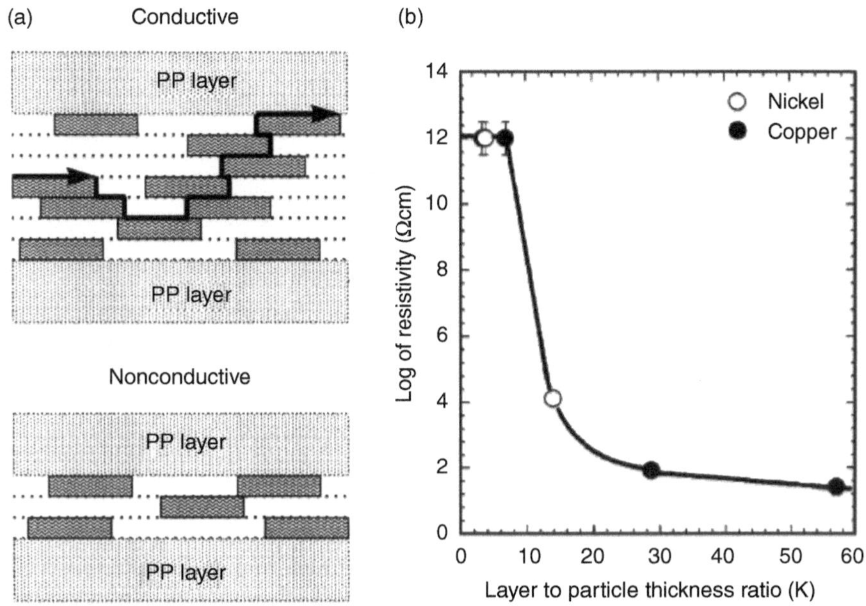

Figure 5.31 (a) Sketch of the particle arrangement in thick and thin filled layers of U–F microlayers. The conductive pathway in the thick layer has been indicated. (b) Logarithm of the volume resistivity versus layer-to-particle thickness ratio (*K*) for copper-filled (solid circles) and nickel-filled (open circles) U–F microlayers.

10% volume fraction for nickel filled layers and 15% volume fraction for copper filled layers are likely to form top-to-bottom-contacts in the thicker layers. As number of layers increases, the decreasing layer thickness approaches the filler particle thickness, and the 2D arrangement of fillers. The required critical loading for percolation threshold in a 2D arrangement increases to 45%, [33] which was significantly higher than the 10 and 15% loadings in the U–F samples investigated. The relationship between the layer thickness and the filler thickness, defined as $K = L/l$, where L is the layer thickness and l is the particle thickness, has been shown in Figure 5.31. The K values can also be dependent on volume fraction as well as filler aspect ratio.

Conclusively, the coextrusion process was successfully used to incorporate metal fillers in the multilayer composites with an increased degree of orientation, which increased the anisotropy in electrical resistivity by two orders of magnitude. Many conductive pathways were also demonstrated in multilayered composites by separating metal-filled layers with unfilled polymer layers.

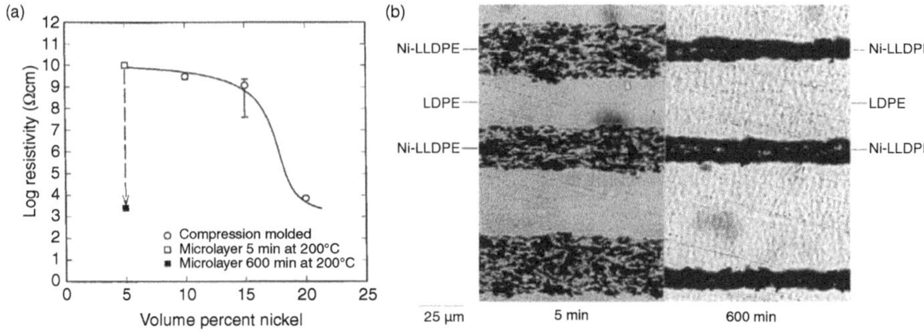

Figure 5.32 (a) Logarithm of the volume resistivity versus filler content of compression-molded nickel-filled LLDPE. In-plane resistivity of the LDPE–Ni–LLDPE microlayer after 5 min and 600 min in the melt at 200°C is indicated. (b) Optical micrographs of the LDPE–Ni–LLDPE microlayer after it was taken to 200°C for the time indicated.

Another approach for concentrating the inorganic particles in multilayered films by thermal annealing was also used to decrease the resistivity below a percolation threshold [34]. The volume resistivity of nickel filled LLDPE resin showed a critical volume fraction of 15% as a percolation threshold. As expected, 32-layered composites of LLDPE and 5% nickel filled LLDPE showed in-plane resistivity of 10^{10} Ωcm, comparable to the compression-molded samples. However, thermal annealing of the composites at 200°C for 10 h resulted in interdiffusion of polymers increasing the concentration of nickel in the filled layer, which showed a decrease of six orders of magnitude in the in-plane resistivity through 10^3–10^4 Ωcm (Figure 5.32a). The thickness of nickel filled LLDPE layer decreased by a factor of two during the annealing process as shown in Figure 5.32b. This example demonstrated an approach to increase the concentration of filler content above the percolation threshold. Furthermore, the samples were expected to demonstrate high anisotropy due to nonconductive properties in cross-plane direction. The detailed interdiffusion of polymers in the multilayer composites and films has been discussed in Chapter 3.

5.3 Summary

Multilayer films with enhanced dielectric properties including breakdown strength, dielectric lifetime, and energy density were fabricated using coextrusion process. The films also showed reduced hysteresis and dielectric losses. These unique properties changes were characteristic of multilayer films of high dielectric constant PVDF-based polymers with high breakdown strength polymers like PC and PET. The breakdown mechanism revealed damages along the layer interfaces by delamination and crazing near the breakdown holes confirmed contribution of interfaces to improved dielectric properties. Further polarization effects in the multilayered films were also investigated by studying orientation related, ionic and interfacial polarizations. These polarizations were affected by various physical properties, such as multilayer film thicknesses, layer thicknesses, interfaces, dielectric contrast, etc.

Furthermore, a strategy to enhance the breakdown properties by modifying the interfaces was also investigated. Tie layer materials with different adhesive properties with the multilayer film components demonstrated the role of interfaces to improve the dielectric properties. When the adhesion of these two components in a multilayer structure was increased using tie layers, the improved breakdown strength and energy density values were observed. Various tie layer materials interacted differently in the multilayer films and produced different interphase fractions in the composite structure. Tie layer materials showed effect on breakdown properties, hysteresis losses, ion migration, and the effective dielectric constant of these layers. An approach of biaxial stretching of PET-based multilayer film was also demonstrated as a route to improve dielectric properties. This film consisted of PVDF-based nanolayers as a part of PET film.

Multilayer coextrusion process was also used to demonstrate fabrication of the metal-filled polymer composites as conducting structures. Effect of copper and nickel fillers on the anisotropic conductivity was investigated as a function of layer thickness in different layer configurations. The metal-filled multilayered composites demonstrated increased resistivity anisotropy by increasing the particle orientation in the filled layers. The layer thickness played a critical role in determining the conductivity properties because the layer thickness of filled layers approached particle thickness, as a result the conductivity properties of these composites were lost.

References

[1] (a) Ho J, Jow TR, Army Research Laboratory: Adelphi, MD, ARL-TR-4880, 2009. (b) Macdougall F, Ennis J, Yang XH, Cooper RA, Gilbert RA, Gilbert JE, Bates JF, Naruo C, Schneider M, Keller N, Joshi S, Jow TR, Ho J, Scozzie CJ, Ten SPS, IEEE Pulsed Power Conference June 28–July 2, Washington, DC, 2009. (c) Clelland I, Price P, Proceedings of the Sixteenth Annual Applied Power Electronics Conference and Exposition, Copenhagen, Denmark, 2001.

[2] (a) W.J. Sarjeant, J. Zirnheld, F.W. MacDoughall, IEEE Trans. Plasm. Sci. 26 (1998) 1368–1392. (b) Q. Tan, P. Irwin, Y. Cao, IEEJ Trans. Fund. Mater. 126 (2001) 1152–1159. (c) I. Husain, Electric and Hybrid Vehicles:

Desin Fundamentals, CRC Press, Boca Raton, FL, 2010. (d) R. Teodorescu, M. Liserre, P. Rodriguez, Grid Converters for Photovoltaic and Wind Power Systems, Wiley, Chichester, UK, 2011.
[3] J. Ho, T.R. Jow, IEEE Trans. Dielec. Elect. Ins. 19 (2012) 990–995.
[4] (a) J. Ho, R. Ramprasad, S. Boggs, IEEE Trans. Dielectr. Electr. Insul. 14 (5) (2007) 1295–1301. (b) H.W. Starweather Jr., P. Avakian, R.R. Matherson Jr., J.J. Fontanella, M.C. Wintersgill, Macromolecules 25 (25) (1992) 6871–6875. (c) W.J. Sarjeant, I.W. Clelland, R.A.P. Price, IEEE 89 (2001) 846–855.
[5] (a) P. Kim, S.C. Jones, P.J. Hotchkiss, J.N. Haddock, B. Kippelen, S.R. Marder, J.W. Perry, Adv. Mat. 19 (7) (2007) 1001–1005. (b) A. Maliakal, H. Katz, P.M. Cotts, S. Subramoney, P. Mirau, J. Am. Chem. Soc. 127 (2005) 14655–14662.
[6] (a) B. Chu, Z. Zhou, K. Ren, B. Neese, M.R. Lin, Q. Wang, F. Bauer, Q.M. Zhang, Science 313 (2006) 334–336. (b) X. Zhou, B. Chu, B. Neese, Q. Zhang, Q. Wang, J. Am. Chem. Soc. 128 (2006) 8120–8121. (c) F. Guan, J. Wang, L. Yang, J.K. Tseng, K. Han, Q. Wang, L. Zhu, Macromolecules 44 (2011) 2190–2199.
[7] J. Lao, D. Jun, Q. Tang, C. Mao, IEEE Trans. Electron. Devices 55 (12) (2008) 3549–3554.
[8] (a) D.P. Agoris, I. Vitellas, O.S. Gefle, S.M. Lebedev, Y.U. Pokholkov, J. Phys. D 34 (24) (2001) 3485. (b) O.S. Gefle, S.M. Lebedev, Y.P. Pokholkov, E. Gockenbach, H. Borsi, J. Phys. D 37 (2004) 2318. (c) S.M. Lebedev, O.S. Gefle, Y.P. Pokholkov, IEEE Trans. Dielectr. Electr. Insul. 12 (2005) 537.
[9] M.A. Wolak, M. Pan, A. Wan, J.S. Shirk, M. Mackey, A. Hiltner, E. Baer, L. Flandin, Appl. Phys. Lett. 92 (2008) 11301–11304.
[10] D.J. Bergman, Phys. Rep. 43 (1978) 377.
[11] M. Mackey, A. Hiltner, E. Baer, L. Flandin, M.A. Wolak, J.S. Shirk, J. Phys. D 42 (2009) 1753304.
[12] M.A. Wolak, A.S. Wan, J.S. Shirk, M. Mackey, Hiltner, E. Baer, J. Appl. Polym. Sci. 123 (2012) 2548–2557.
[13] Z. Zhou, M. Mackey, Carr, L. Zhu, L. Flandin, E. Baer, J. Polym. Sci. B 50 (14) (2012) 993–1003.
[14] M.H. Lean, A. Wolak, M. Mackey, E. Baer, IEEE Trans. Dieler. Electr. Insul. 21 (2) (2014) 800–808.
[15] Zhou Z, Mackey M, Yin K, Zhu L, Schuele D, Flandin L, Baer E, J. Appl. Polym. Sci. 131(4), doi: 10.1002/app.39877.
[16] C. Reed, S. Cichanowski, IEEE Trans. Dielectr. Electr. Insul. 1 (1994) 904–922.
[17] M. Rabuffi, G. Picci, IEEE Trans. Plasma Sci. 30 (2002) 193901942.
[18] (a) L. Zhu, Q. Wang, Macromolecules 45 (2012) 2937–2954. (b) L. Zhu, J. Phys. Chem. Lett. 5 (2014) 3677–3687.
[19] M. Mackey, D.E. Schuele, L. Zhu, L. Flandin, M.A. Wolak, J.S. Shirk, A. Hiltner, E. Baer, Macromolecules 45 (2012) 1954–1962.

[20] H.S. Nalwa, Ferroelectric Polymers: Chemistry, Physics, and Applications, Marcel Dekker, Inc, New York, 1995.
[21] F. Guan, J. Pan, J. Wang, Q. Wang, L. Zhu, Macromolecules 43 (2010) 384–392.
[22] (a) N. Rashmi, P.K.C. Pillai, J. Mater. Sci. 22 (1987) 2006–2010. (b) E. Tuncer, M. Wegener, R. Gerhard-Multhaupt, J. Non-Cryst. Solids 351 (2005) 2917–2921.
[23] S. Uemura, J. Polym. Sci. Polym. Phys. Ed. 10 (1972) 2155–2166.
[24] J.K. Tseng, S. Tang, Z. Zhou, M. Mackey, J.M. Carr, R. Mu, L. Flandin, D.E. Schuele, E. Baer, L. Zhu, Polymer 55 (2014) 8–14.
[25] Verkhovskaya KA, Saidov AC, Vannikov AV, Proceedings of the SPIE Xerographic Photoreceptors and Organic Photoreactive Materials IV, 3471, 117–120, 1998.
[26] Z. Zhou, J. Carr, M. Mackey, K. Yin, D. Shuele, L. Zhu, E. Baer, J. Polym. Sci. Part B Polym. Phys. 51 (2013) 978–991.
[27] J. Mijovic, J.W. Sy, T.K. Kwei, Macromolecules 30 (10) (1997) 3042–3050.
[28] S. Kang, Y. Park, I. Baer, K. Kim, H. Kim, S. Bauer, E. Thomas, C. Park, Adv. Funct. Mater. 19 (2009) 2812–2818.
[29] J.M. Carr, M. Mackey, L. Flandin, D. Schuele, L. Zhu, E. Baer, J. Polym. Sci. Part B Polym. Phys. 51 (2013) 882–896.
[30] (a)A.T. Ponomarenko, V.G. Shevchenko, N.S. Enikolopyan, Adv. Polym. Sci. 125 (1996). (b)D.M. Bigg, in: S.K. Bhattacharya (Ed.), Metal-Filled Polymers, Mercel Dekker, New York, 1986, pp. 165.
[31] S. Nazarenko, A. Hiltner, E. Baer, J. Mat. Sci. 34 (1999) 1461–1470.
[32] (a) J. Gurlang, Trans. Metal. Soc. AIME 236 (1966) 642. (b) L. Nikodemo, L. Nikolais, G. Romeao, E. Scafora, Polym. Eng. Sci 18 (1978) 293.
[33] (a) G.E. Pike, C.H. Seager, Phys. Rev. B 10 (1994) 1421. (b) A.T. Ponomarekno, V.G. Shevchenko, T. Tchmutin, A.A. Ovchinnikov, N.S. Enikolopyan, Polymer Composites Ed. Sedlacek B, Walter de Gruyter and Co, Berlin, New York, 1986, p. 67.
[34] S. Nazarenko, M. Dennison, T. Schuman, E.V. Stepanov, A. Hiltner, E. Baer, J. Appl. Polym. Sci. 73 (1999) 2877–2885.

6 Novel Multilayered Structures and Applications

6.1 Introduction

Multilayered films and composites containing hundreds of polymer layers demonstrated unique properties as discussed in earlier chapters of this book. Various coextrusion processing techniques and innovations with different feedblocks, layer multiplying dies, and annular dies are discussed in Chapter 2. As demonstrated in many multilayered film systems, the confinement of polymers at micro- and nanoscale affects the structure and properties of these films. The films and composites consisting of a few tens to a hundreds of layers demonstrated dramatic effects on the gas transport, mechanical, optical, electrical, and many other properties as discussed in previous chapters. For example, coextruded films consisting of PEO or PCL nanolayers demonstrated improved gas barrier properties, due to formation of oriented lamellae under nanoconfinement. Many other polymers confined under suitable crystallization conditions also showed improvement in the gas and water barrier properties. The structure–property analysis of coextruded structures in these examples revealed that, crystallization behavior of the polymers is significantly altered as layer thicknesses approached the radius of gyration of polymer molecules. In addition, the effects on adhesive and mechanical properties were attributed to changes in the interphase formation and interdiffusion at the layer interfaces. In fact, the coextruded structures with thousands of layers demonstrated formation of entirely interfacial materials with immiscible polymers. Other important applications of the multilayered films such as optically reflective films, and 3D data storage composites are discussed in Chapter 4. Furthermore, the recent innovation included development of multilayered dielectric films with enhanced energy storage capabilities, as compared to conventional BOPP and PET films for capacitor applications. It is obvious that the multilayered films and composites have several advantages over conventional polymers. In this chapter, many other novel multilayer structures with potential applications in blends and composites, membrane filtration, foams, porous materials, optical films, shape memory composites, and fibers are reviewed.

Manufacturing and Novel Applications of Multilayer Polymer Films
http://dx.doi.org/10.1016/B978-0-323-37125-4.00006-X

6.2 Blends and Composites

Polymer blending is a commonly used approach in product development to achieve synergistic effects by combining two or more constituents with different properties. The intrinsic properties of these constituents have a significant effect on the resulting blend properties. In addition to the blend components, many other critical parameters controlling final properties include compositions, formulations, processing parameters, etc. Different concepts and principles of blending technology are thoroughly discussed in many publications and books [1]. Because the coextrusion process combines two or more polymers, the layered structures can be diced or pelletized, and used as blend materials in postprocessing techniques such as molding.

6.2.1 Innovative Blends

The coextrusion process typically produces films, sheets, and annular profiles, which can be further processed below certain maximum temperatures to produce bags, packages, or thermoformed products without disturbing the layered structure morphology. In contrast, when coextruded structures were heated above melting temperatures of component polymers, thermal break-up of layered structures resulted in a blend-like morphology. Interestingly, the coextrusion technology is known as a "forced assembly" process, implying that the layered structures may not be stable when heated into the melt state. As demonstrated in many examples, when layered films were heated above the melting temperature of component polymers, irreversible layer break-up resulted in the creation of melt droplets of one polymer dispersed into other polymer matrices. This phenomenon of interfacially driven layer break-up prevents multilayer film structure integrity above the melting temperatures of component polymers. Interestingly, the resulting dispersion of polymer droplets from thermal layer break-up resembled the blend structures obtained by conventional blending techniques. An example of the thermal break-up of 40 nm HDPE nanolayers into dispersed droplets with increasing temperature is shown in Figure 6.1 [2]. The layer instability in the continuous HDPE layers was noticed at 138°C, slightly above the HDPE melting temperature of 134°C. The layer break-up and formation of polymer droplets became pronounced with increasing temperature. Moreover, the droplet size of dispersed polymer was dependent on individual layer thickness in the layered structure [3]. For example, when the HDPE layer thicknesses were only a few nanometers in a 257-layered 5/95 (vol/vol) HDPE/PS multilayer film system, most of the dispersed HDPE polymer droplets were produced in submicrometer-sized range. Fractionated crystallization of these polymer droplets was

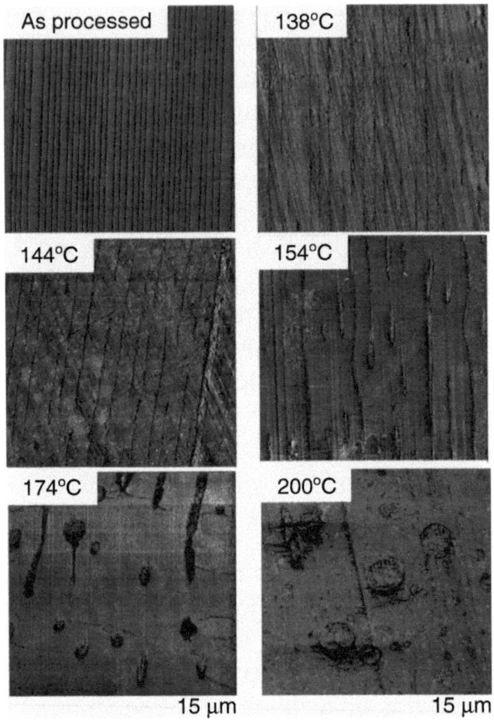

Figure 6.1 AFM phase images showing the cross-section of HDPE/PS films after heating to the temperature indicated. The nominal HDPE layer thickness was 40 nm.

successfully used to investigate homogeneous and heterogeneous nucleation in polymers [4]. This approach was further extended to incorporate additives and nucleating agents in the polymer droplets to study the nature of heterogeneous nucleation. It is obvious that the injection or compression molded specimens from A/B type multilayered structures will resemble conventional polymer blends when the pelletized multilayered structures are heated above the layer break-up temperature of the constituent polymers. The relative composition of two polymers will play an important role in controlling the polymer droplet size. A cocontinuous morphology is also possible in thermally treated 50/50 volume composition films [2].

In contrast, when A/B type multilayered structures were pelletized and injection molded by selectively melting only one of the component polymers, high aspect ratio microplatelets of the second polymer were successfully produced. The microplatelets resembled the "oriented morphology" of dispersed polymer phase in a polymer matrix, which offers unique advantages to improve the blend properties. For example, a blend of high gas barrier polymers such as polyamides or ethylene vinyl alcohol copolymers (EVOHs) with a polyolefinic material, can benefit from oriented morphology of the barrier polymer, as it creates a tortuous path for gas diffusion. Ideally, the multilayer coextrusion process can create highly ordered laminar morphology for barrier improvement, as demonstrated in numerous examples discussed earlier in this book. However, as the coextrusion process approach is limited to film, sheet, annular profiles or thermoformed products; the coextruded pellets (or diced multilayered films) can be used as blends in other molding techniques at suitable processing temperatures to achieve microplatelet-based oriented morphology. The high aspect ratio microplatelets of polymers have demonstrated many advantages in improving barrier properties, mechanical properties, and filler dispersion [5,6]. By varying the number of layers and individual layer thicknesses down to few 10s of nanometers, it is possible to produce high aspect ratio micro- or nanoplatelet blends. An example of this approach includes injection-molded plaques creating polyamide-66 (PA66) microplatelets dispersed in polypropylene (PP) matrix, to improve the gas barrier properties. In the first step, the layered composites of PP/(PA66) were fabricated with PP-*g*-MA tie layer, which improved the adhesion between incompatible PP and PA66 polymers. PP was selected for its excellent water barrier properties, while PA66 was used for its high gas barrier properties. Multilayered films with 33, 257, and 2049 number of layers, were produced with 10% PP-*g*-MA content [5]. As polymers selected for processing demonstrated large viscosity mismatch, a broad layer thickness distribution was observed in these film samples. The effect of viscosity on layer structure is discussed in Chapter 2. The layered films were produced at a processing temperature of 280°C, which was significantly higher than the PP (T_m = 163°C) and PA66 (T_m = 262°C) melting temperatures. The multilayered films were diced and injection molded at 180°C, an intermediate temperature between PP and PA66 melting points. As expected, the difference in the melting points and intermediate processing temperatures selectively melted PP layers, while PA66 layers remained solid and dispersed as high aspect ratio microplatelets in PP matrix. The morphology analysis of injection molded parts showed highly oriented PA66 platelets parallel to the surface of molded parts, while the core sections showed coiled platelets with preferential orientation perpendicular to the injection direction, due to complex flow during injection molding process. The structure development in injection molded specimens as a function of number of layers is shown in Figure 6.2. As the curling of the layers occurred predominantly near the surface, the differences in the morphological features were apparent in longitudinal direction. Although, the PA66 microplatelets remained parallel to the injection direction, pronounced layer curling was observed

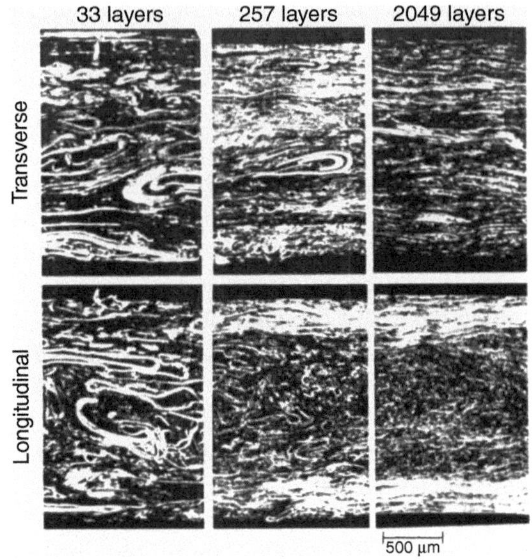

Figure 6.2 Optical micrographs of transverse and longitudinal sections through the thickness of injection molded microlayers. Microlayers with 33,257 and 2,049 layers were injection molded at 180°C.

with increasing number of layers. The optical micrographs of sections of injection-molded plaques in transverse and longitudinal direction show the effect of a number of layers on the morphology. Additionally, the PA66 layer lateral dimensions were an order of magnitude smaller than the initial pellet size, suggesting layer breakup during injection molding. The morphological features remained unchanged at higher processing temperatures of 240°C, close to PA66 melting onset. The injection molded plaques with oriented PA66 layers showed a four- to fivefold improvement in the barrier properties over conventional melt blends of PP and PA66. An example demonstrating improved barrier properties is discussed in Chapter 3.

This approach was further extended to produce polymer pellets with 4096 alternating layers of polysulfone (PSU) and ethylene–octene (EO) copolymers [6]. The extruded structure was produced as melt strands, which were pelletized into granular form for compression and injection molding trials. The difference between glass transition temperature of PSU (T_g = 186°C) and the EO melting temperature of 130°C allowed a processing of the layered pellets above EO melting and below PSU T_g to produce dispersed PSU microplatelets. The injection and compression molded plaques revealed that PSU layers remained intact without significant aggregation between 130°C and 180°C processing temperature. The injection-molded samples with 10 wt% PSU platelets showed improved tensile and flexural properties as compared to conventional PSU–EO blends.

It is obvious from these examples that relative melting temperatures of polymers and molding temperatures play significant roles in creating polymer droplets or microplatelets containing composite. This technique also demonstrates a novel approach in creating polymer composites with property improvements beyond conventional blending processes [5].

6.2.2 Filled Composites

The coextrusion processing approach was further utilized to fabricate layered composites with alternating layers of filled and nonfilled polymers. For example, multilayered strands of PP and inorganic fiber-filled PP (FPP) were extruded to create 64-layered structures with 50/50 volume composition of the two components and individual layer thicknesses of 1.2 μm. Magnesium-based inorganic whisker, Hyperform HPR-803, was blended with PP to create a FPP blend in this example. SEM images revealed dispersion of fibers in alternating layers and orientation in the extrusion direction, as shown in Figure 6.3a [6]. These composites were used as pellets for injection molding trials to produce plaques, which showed a twofold increase in flexural modulus as compared to melt blend composites.

Another example demonstrated composite structures with alternating layers of phosphate glass-filled PP-*g*-MA with neat PP-*g*-MA, Figure 6.3b. The inorganic phosphate glasses are blended with organic polymers to improve barrier properties [7]. Because the phosphate glass T_g was 105°C, the biaxial stretching of these films created highly oriented glass platelets in the layers exhibiting brick wall type microstructure with significantly improved barrier properties as discussed in Chapter 3. In addition, dispersed phosphate glass platelets also improved the composite modulus

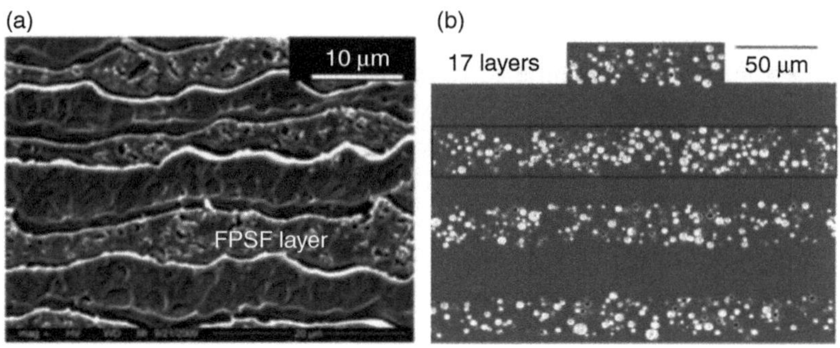

Figure 6.3 (a) SEM image of the cross-section of fiber embedded PSU and PP multilayer composite and (b) an SEM image cross-section of multilayered film with 10 vol% phosphate glass [6].

without sacrificing film ductility. This structure in fact resembled natural bio-composites with alternating organic and inorganic layers. Other examples of this technology demonstrated coextruded films of PMMA with graphene nanoplatelets-filled PS or PMMA to achieve planar orientation of graphene [8]. The multilayer coextrusion process was also used to achieve exfoliation of graphene from graphite [9].

Similar to phosphate glass-filled composites, in other examples, the two base polymers constituting coextruded structures were retained constant and inorganic fillers were present only in alternate layers to create a brick-wall type structure. Increase in the concentration of inorganic fillers by interdiffusion was achieved in these samples. As discussed in Chapter 3, multilayered films with alternating layers of LLDPE-, TiO_2- or Nickel-filled LLDPE were coextruded to create hybrid composites. The thermal annealing of these films moved the layer boundaries and increased the concentration of fillers in alternating layers [6,10] (Figure 6.3).

6.3 Multilayered Foams

Polymer foams are widely used in many applications such as thermal insulation, food, and beverage packaging, construction, electronics, sound attenuation, filtration, etc. Some commonly used polymer foams fabricated using polystyrene (PS), polyurethane (PU), polyethylene (PE), PP, and polyolefinic copolymers offer different properties [11]. For example, PS foam, one of the first developed and most commonly used foam material, suffers from inferior mechanical properties such as low fracture strain. Another foam material, PU, can be fabricated into different formulations, but it is expensive than PS or polyolefinic foams. Development of PP and PE foams in recent years, have been successful due to their low cost and improved mechanical properties. Although recent development of EO copolymer-based foams resulted in elastic and resilient foams, the elastomeric foams showed relatively large variation in cell sizes due to diffusion of blowing agent during the foaming process [12]. Additionally, at low bulk densities of foams, the mechanical properties deteriorate significantly. Types of foam raw materials, processing and microstructures, types of foams, and different applications are discussed in detail in other publications [11]. Novel approaches of multilayer coextrusion processing demonstrated microcellular-polyolefinic structures with alternating foam and film layers ("foam/film" composite) and cell size control. Furthermore, the cell orientation in the foam/film composites enhanced toughness, tear resistance, and puncture resistance [13].

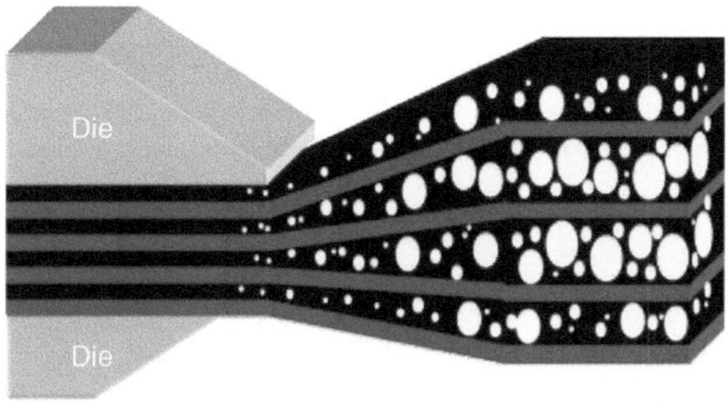

Figure 6.4 Schematic of foam/film multilayer coextrusion process with different stages of foaming processes in a film.

Different types of multilayer foam/film structures investigated include PP, PE, ethyelene–octene copolymer, cross-linked polyolefin elastomers, and polyolefin blends [12]. Coextrusion process created alternating layers of two polymers – first polymer contained a chemical blowing agent and a nucleating agent necessary for foaming, while second polymer formed film layers separating the foam layers. The foam/film production process is schematically shown in Figure 6.4 [14]. The presence of a chemical blowing agent and nucleating agent resulted in nucleation, bubble growth, and stabilization of bubbles. During processing, the chemical blowing agent decomposes to release the gas. The amount of gas released does not nucleate if it's below the solubility limit in the polymer. The dissolved gas nucleates and forms gas bubbles as the polymer melt exits the cast die, resulting in pressure drop. The bubbles grow with further gas diffusion followed by bubble stabilization upon sheet cooling. The film layer stability is dependent on bubble growth stress and melt strength of the film layer, creating a foam/film structure. Other parameters affecting the layer stability and cell size are polymer type, relative melt strength of the coextruded polymers, blowing agent concentration, nucleating agent concentration, and viscosity among few. Examples of the coextruded multilayered foam/film structures created using different types of polymers are discussed later [14].

6.3.1 Polypropylene-Based Foam/Films

Improved chemical resistance, temperature stability, and mechanical properties can be achieved in PP foams as compared to PS foams. However, the cell size in commercial PP foams is not uniform due to low solubility of gases and nonhomogeneous nucleation in PP [15]. Development of PP-based foam/film structures with cell size control as a function of number of layers was demonstrated in the layered composites. A two-component

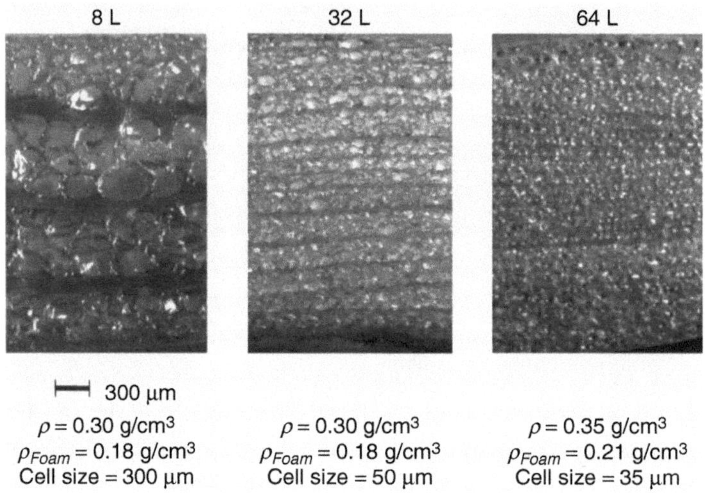

Figure 6.5 Effect of number of layers on foam/film structure for PP foam/PP film (50/50).

coextrusion processing was used to produce PP-based foam/film structures with 50/50 (vol/vol) as shown in Figure 6.5. Azodicarbonamide (Azo) or SAFOAM chemical agent, and magnesium oxide nucleating agent were used to foam PP layers. As the number of layers increased from 8 to 64, the cell size decreased from 300 μm to 35 μm, and the cell density increased. It is evident that the presence of a confining PP film layer prevented the cell coalescence and/or increased the cell nucleation. The cell size decreased 10-fold with increasing numbers of layers, without significant impact on the foam density. The calculated foam layer density did not change significantly with number of layers and it remained constant at 0.18–0.21 g/cm^3, while the overall density of the composite structure varied between 0.30 g/cm^3 and 0.35 g/cm^3. Interestingly, 35 μm cell size and the foam layer density of 0.18 g/cm^3 were comparable with the cork cell density of 30 μm and density of 0.18 g/cm^3. Furthermore, the effect of the composition of PP foam and PP film in the 32-layered composite showed no effect on the overall density as shown in Table 6.1. However, the cell size decreased significantly from 80 ± 10 μm in 30/70 foam/film composite to 45 ± 15 μm in 80/20 foam/film composite. The foaming efficiency of 90% in all samples was determined by calculating the amount of the CO_2 gas in the foam layer and the CO_2 estimated from the thermogravimetric analysis of the blowing agent, as discussed in Ref. [14].

The effect of the number of layers on stress–strain behavior of foam/film composites in compression and tension are shown in Figure 6.6. Increased

Table 6.1 Effect of Foam/Film Composition on Overall Density, Foaming Efficiency, and Cell Size

PP Foam/PP Film Compositions	Overall Densities (g/cm³)	Foam Densities (g/cm³)	Foaming Efficiencies (%)	Cell Sizes (μm)
100/0	0.17	0.17	90	300 ± 50
80/20	0.29	0.16	92	45 ± 15
50/50	0.30	0.18	88	50 ± 10
30/70	0.31	0.16	92	80 ± 10

moduli were observed in both compression and tensile testing of the samples. During compression, cell bending and cell boundaries stretching occurred at low strains. With increasing strain, the stress remained unchanged from 5% to 65% as the cell collapsed, while subsequent increase in the stress indicated densification process. The tensile behavior shows strain at break of 12% for samples with variable number of layers. To improve the fracture strain properties of the composite, multilayered foam/film systems with PP foam layers and EO copolymer film layers were produced. In this example, foam layer contained a blend of PP and PE. Layered foam/film composite of PP-blend/EO showed a strain at break of 375%, which was significantly higher than tensile at break of 12% in PP foam/PP film composite system. The composite structures showed cell size in the range of 60–80 μm in 32-layered composites for compositions ranging from 80/20 to 50/50 (PP blend foam/EO film). The overall foam density remained constant at 0.30 g/cm³, similar to PP foam/PP film system [14].

6.3.2 Ethylene–Octene Copolymer-Based Multilayer Foams

Another example showed foamed EO layers confined by LDPE layers. When EO copolymer was foamed as a single layer film, most of the gas readily diffused through the structure during processing, prior to cell wall formation with foaming efficiency of only 3%. In contrast, when LDPE was used as an alternating layer, the gas remained confined to EO layer increasing the foaming efficiency. As the LDPE layer thickness increased, foaming efficiency increased from 3% in EO control to 21% in 70/30 (EO/LDPE) composition to 50% in 50/50 (EO/LDPE) composition. However, this foaming efficiency was not as high as 90%, as observed for PP-based foam/film systems.

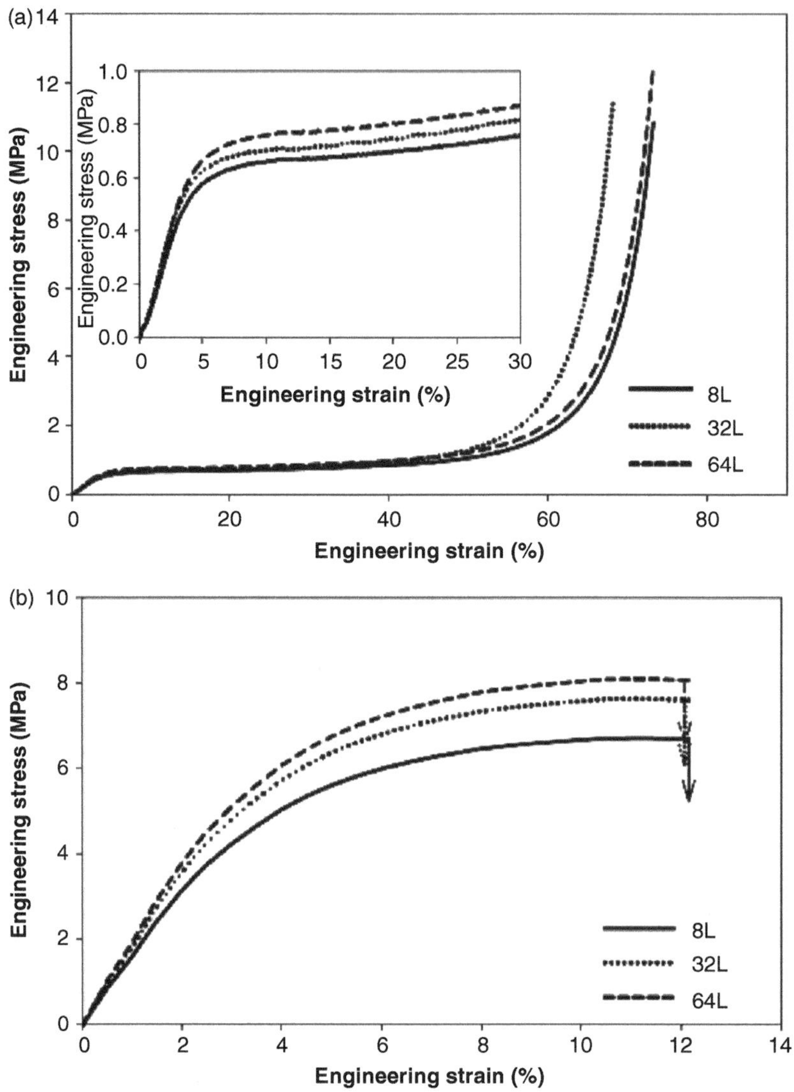

Figure 6.6 Effect of number of layers on stress–strain behavior of PP foam (1.75% Azo)/PP film (50/50). (a) Compression and (b) tension.

Foam/film composites of EO elastomer foam alternating with blends of EO and LLDPE have demonstrated use as sound absorbing properties [16]. It was found that the sound absorption coefficient increased gradually in single layer foam composite in the range of 500–2300 Hz. In contrast, 8-layered foam/film composites, exhibited two maxima at 1550 and

2000 Hz frequencies, while 32-layered composite showed maxima at 950 and 1500 Hz. In addition, these layered foam/film composites showed improved tensile strength and tensile modulus in as compared to single foam layer composites.

6.4 Porous Composites

6.4.1 Multilayered Gas Separation Membranes

Gas separation membranes are used in diverse applications such as gas purification, separation of byproducts, modified atmosphere packaging (MAP) applications [17]. Ceramic and alloy-based membrane technologies have historically been used for their excellent chemical resistance and stability. The polymer-based technologies use polymer separation membranes supported on porous substrates, which provide mechanical durability and stability. Fabrication of ultrathin separation membranes is typically achieved by solution casting or interfacial polymerization, while the porous support is fabricated by phase inversion techniques. Porous support materials can also be nonwoven mats produced by various airlaid, wetlaid, electrospinning, or melt spinning techniques [18]. It is obvious that multiple energy intensive fabrication steps and significant handling of organic solvents are necessary to create polymer separation membranes with good mechanical properties. To address these challenges, an approach of coextrusion processing was explored to create membranes using a two-polymer multilayer film system, where the first material was selected as a gas separation membrane polymer (highly selective polymer) and the second polymer created porous structures [19].

A poly(ether-block-amide) (PEBA) polymer with 86 mol% polytetramethyl oxide (PTMO) and 14 mol% polyamide-12 (PA-12) was used for its high CO_2/O_2 selectively and high gas permeability. Two approaches of creating porous layers with alternating gas separation layers were explored. In the first approach, PEBA was coextruded with PP blends containing $CaCO_3$, which created porous PP layers during uniaxial stretching of the layer composite. In the second approach, a multilayered film of PEBA with β-phase PP, achieved by blending it with a β-phase nucleating agent quinacridonequinone (QQ), was extruded. The β-nucleated PP transformed into α-phase during stretching and created porous features due to void formation between the lamellae during this transformation.

The multilayered structures, PEBA/(PP + $CaCO_3$) and PEBA/β-PP were produced in 30/70 and 10/90 volume composition films with varying number of layers from 2 to 17. An example of a three-layered extruded

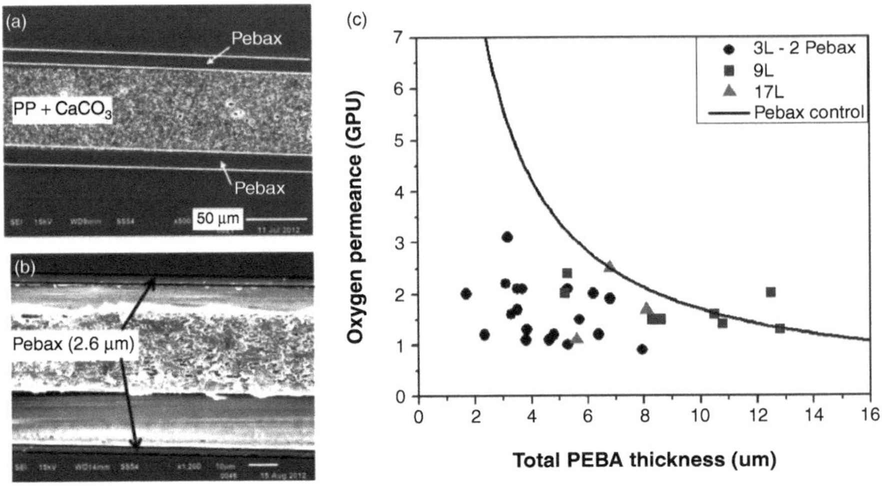

Figure 6.7 (a) As-extruded symmetric three layer films of 30/70 volume composition PEBA/PP + $CaCO_3$; (b) uniaxiallyoriented PEBA/PP + $CaCO_3$ films; (c) comparison of oxygen permeance for PEBA/PP + $CaCO_3$ multilayed (3-, 9-, and 17-layered) membranes.

film with PEBA as outside layers and PP + $CaCO_3$ nonporous core layer is shown in Figure 6.7a. Film orientation by uniaxial stretching induced porosity in PP layers at $CaCO_3$ sites, while reducing PEBA layer thickness, Figure 6.7b. As the film orientation induced porosity, the oxygen, and CO_2 flux increased due to porous PP layers. The membrane composites showed the CO_2/O_2 selectivity of at least seven. The effect of number of layers on the oxygen permeance values is shown in Figure 6.7. The oxygen permeance in the layered composites below 6 μm PEBA layer thickness was less than the PEBA control. Postorientation annealing techniques were used to improve the oxygen permeance of the membranes. For example, annealing PEBA/PP + $CaCO_3$ films above PEBA melting temperature increased the oxygen and CO_2 flux by 100% and showed selective improvement from 8.7 to 9.6. The permeability values of the multilayered membrane structure were not significantly different in porous membranes created using $CaCO_3$ filler or beta phase transformation. Figure 6.8 shows the oxygen permeability and the CO_2/O_2 selectivity of the PEBA/β-PP membranes [20]. The multilayered membranes showed oxygen permeability values close to 5 GPU (GPU = Barrer/μm), which was significantly less than 30 GPU required for a standard MAP package [21]. The 3-layered structures showed highest oxygen permeabilities and the permeance than the 17-layered films possibly due to reduced porosity at higher number of layers. Further structural evaluation revealed the

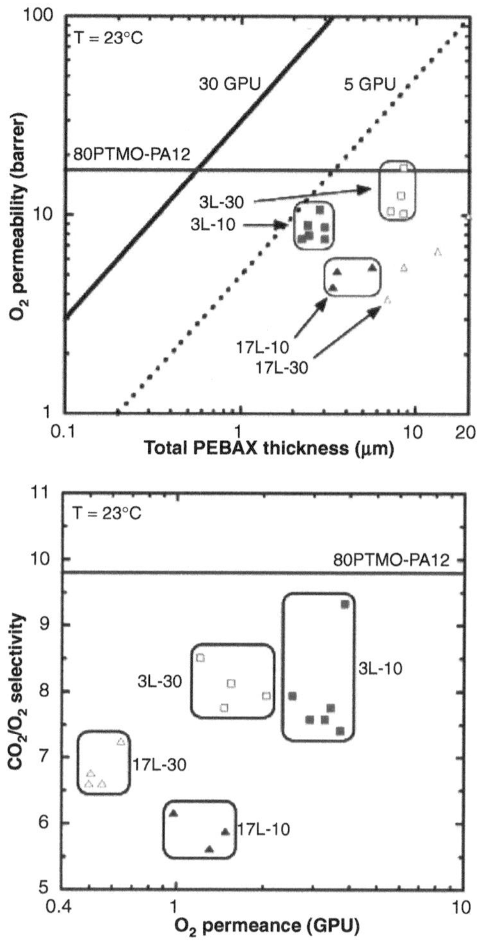

Figure 6.8 O_2 permeability of multilayered films as a function of total PEBAX thickness (top) and multilayered film CO_2/O_2 selectivity as a function of film permeance (bottom). The permeability and selectivity of a single layer 80PTMO-PA12 (Pebax) film are shown. Results for 3L-10, 3L-30, 17L-10, and 17L-30 films are presented. (80PTMO-PA12 contains 80 wt% of poly(tetramethyloxide) and PA12 is a polyamide block. 3L-10 represents three layers and 10% PEBAX, 3L-represents 3-layered films with 30% PEBAX, 17L-10 represents 17-layered film with 10% PEBAX, and 17L-30 represents 17-layered film with 30% PEBAX.)

inconsistent pore formation in the layers resulting into higher selectivities but lower permeance values. The modeling results revealed the nonporous areas in PP layers negatively impacting the transport properties. For PEBA/β-PP multilayered composites, the CO_2/O_2 selectivity varied between six and nine in these multilayered films.

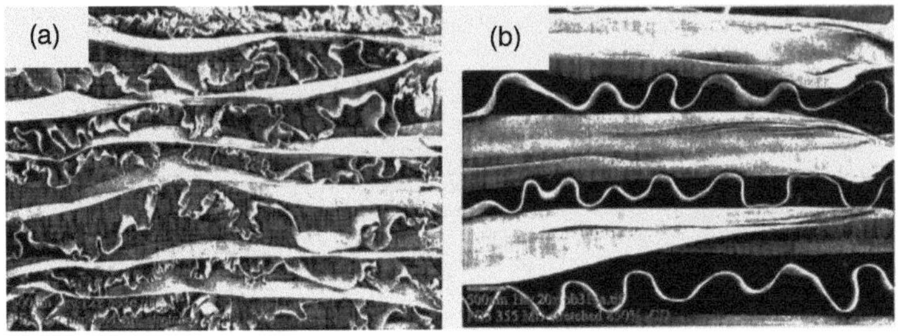

Figure 6.9 SEM micrograph showing cross-sectional view in machine direction of a corrugated microlayer film. At (a) 12× magnification and (b) 20× magnification.

The coextrusion technology demonstrated a solvent-free approach to create supported polymer membranes. Although, this technology is in early research stage, it offers a unique approach in creating MAPs. Process optimization to create porous alternating layers will be useful toward producing commercial membranes [19,20].

6.4.2 Corrugated Layers as Porous Materials

Stretched multilayered films with alternating plastic and elastic layers created corrugated structures as breathable films for use as absorbable components. Buckling of the plastic layer during the recovery of multilayered films consisting of alternating plastic/elastomer layers formed corrugated structures as shown in Figure 6.9 [22]. As discussed in many examples in Ref. [21], coextruded multilayer film contained a filled polyolefin material as plastic layers, while elastomeric layers was selected from wide range of materials such as PEBAX, PUs, etc. Early investigation on the corrugation, buckling and recovery, was carried out in three-layered elastomer/plastic/elastomer films using SEPS block copolymer elastomer. An elastic model was developed to describe the critical wavelength, critical compression strain, and stress to understand buckling. The linear dependence of the critical wavelength of the buckled layer on the plastic layer thickness was observed in the films. This understanding can be used to develop multilayer film systems with corrugated plastic layers separated by elastomeric layers as porous materials. In fact, this approach was extensively investigated to create breathable films with possible uses as absorbable components in personal care products [23]. Furthermore, it was also suggested that the particulate fillers such as $CaCO_3$ can be used to create the pores for improving breathability of the films [22].

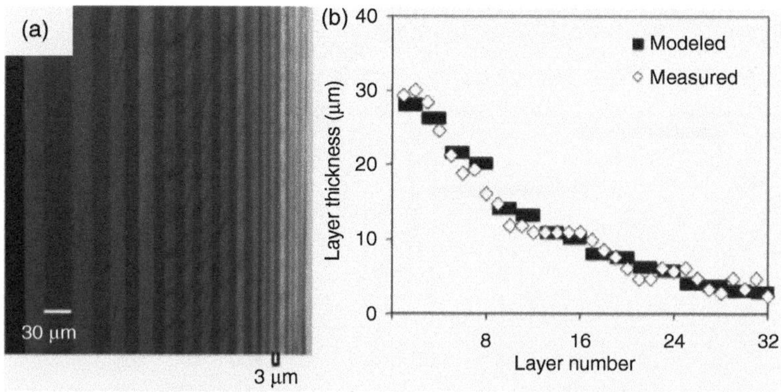

Figure 6.10 Cross-section optical micrograph of a 32-layered PS/PMMA gradient layer thickness film with a 10× difference between the thickest and thinnest layers.

6.5 Gradient Multilayered Films

Multilayered structures with gradient layer structures are commonly found in nature exhibiting vibrant colors, optical reflections in many organisms [24]. Many optical phenomena associated with the gradient-layered composites are discussed in Chapter 4. This section discusses coextrusion process for producing gradient layers and some relevant applications. Commercially the gradient layer thickness distribution is achieved by using custom designed multilayer feedblocks with fixed number of layers and layer thickness distribution [25]. In a recent approach, development of the uneven split multipliers demonstrated fabrication of gradient layer polymer films with processing flexibility. An example of a 32-layered gradient film with alternating PS and PMMA layers and a 10×-layered thickness gradient is shown in Figure 6.10. The overall composition of the two components was maintained at 50/50 (vol/vol). The processing involved layer multiplication of 50/50 composition of PS/PMMA followed by four multiplication steps with increasing offset split ratio to 51/49, 52/48, 54/46, and 58/42. The measured layer thickness distribution matched closely with the modeled layer thickness distribution for uneven split multipliers. This new advancement offers the process flexibility to rearrange the multiplier locations in the extrusion process or use "even" or "uneven" melt splitting elements (multipliers) in any order to create custom designed layer thickness distributions within the multilayer film [26].

6.5.1 Optically Reflective Gradient Films

Ability of the layered films to combine the polymers with different refractive indices to produce optically reflective films is discussed in

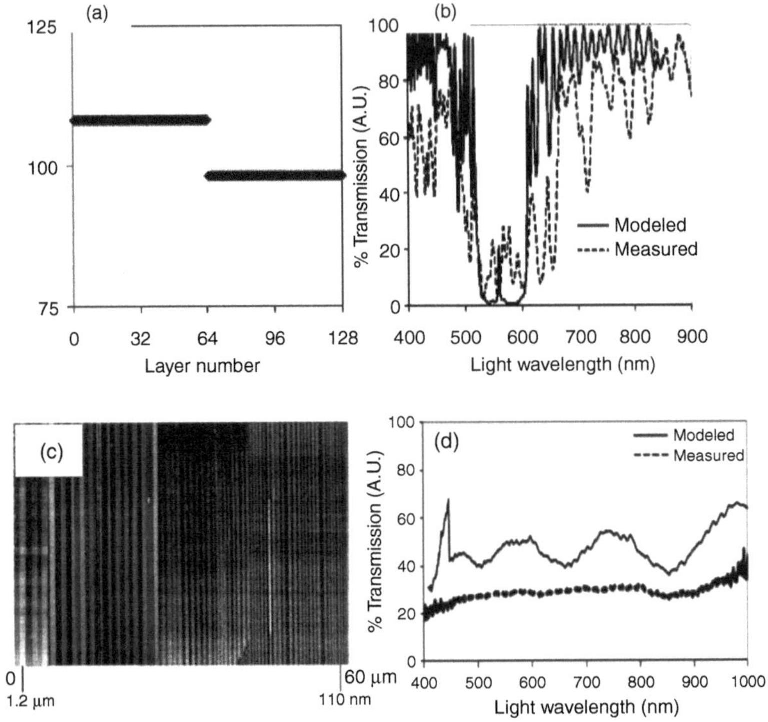

Figure 6.11 (a) A 128 step-layered thickness distribution; (b) resultant comparison between the analytically modeled and the experimentally measured optical reflection spectra from a 128 multilayer PS/PMMA film; (c) cross-section atomic force microscopy image of 128 gradient layer film with a 10× layer thickness distribution produced by coextrusion through a series of two even and four uneven multiplier dies; and (d) comparison of analytical prediction and experimentally measured 600+ nm wide reflection band from 128 gradient layer THV/PS multilayer film.

Chapter 4. Figure 6.11a demonstrates a 128-layered composite film with 50/50 PS/PMMA (vol/vol) composition with two layer thickness distributions, 0.5-mil thickness and a refractive index difference (Δn) of 0.08 [26]. As the film constituted two sets of layer thicknesses, a widened optical reflection peak from 520 nm to 680 nm was observed. The broadened optical reflection was 3× wider than the uniform layer thickness PS/PMMA (50/50) films [27]. Polymer films with tunable optical bandgaps of 100–200 nm can be designed with this flexible process. In another example, a 128-layered film of PS/THV (50/50 v/v) with a Δn of 0.19 was coextruded using two even split and four uneven split multipliers to produce a wide band optical reflection. The coextruded film showed layer thickness distribution from 1.2 μm to 110 nm, Figure 6.11c. This

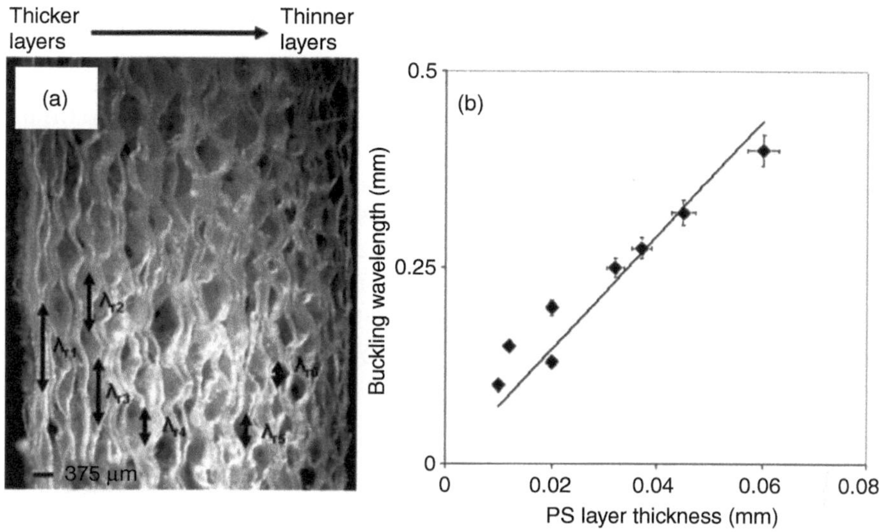

Figure 6.12 (a) A 128-gradient-layered 50/50 PS/THV foam-delaminated film exhibiting layer thickness dependent layer buckling. (b) Effect of PS layer buckling wavelength with layer thickness in 128-gradient-layered 50/50 PS/THV foam-delaminated film.

gradient structure exhibited a broad optical reflection spectrum, greater than 600 nm, across the entire visible light spectrum with the measured optical reflection intensity of 20%, Figure 6.11d. Additionally, the modeling results revealed that the coextruded film could be utilized above 1000 nm (near IR) and below 400 nm (UV) [26].

6.5.2 Gradient Foams

As discussed in Section 6.2, the cell size of foamed polymers is dependent on the starting layer thickness in a film/foam composite. A PS/THV gradient film with a 10× difference in the layer thickness and a chemical blowing agent in PS layers, demonstrated the gradient in the cell size as shown in Figure 6.12 [26]. As the gradient film contained alternating plastic/elastic layers, the foaming phenomena induced layer buckling in the plastic layer. The structure also resembles corrugated structure discussed in Section 6.3. The gradient films can be used to create the corrugated structures by selecting appropriate materials (plastic/elastic system) and straining conditions. Gradient films with controlled mechanical properties were also designed transition from brittle crazing to ductile shear banding. A detailed summary of the research is discussed in Chapter 3 [26].

6.6 Shape Memory Polymers

The shape memory polymers (SMPs) represent a class of materials capable of shape change and shape recovery under external physical stimuli such as strain, temperature, moisture, pH, etc. SMPs transition between permanent and temporary shapes using external stimulus. They can be used in a wide range of applications such as medical devices, sensors and actuators, self-deployable structures, textiles, etc. [28]. Several publications and review papers have discussed synthesis, structure–property relationships and applications of SMPs in detail [29]. In summary, the basic requirements to achieve shape memory properties, shape fixing and shape recovery, are: (a) physical or covalent cross-links with reversible extensions; and (b) network chains between the cross-links with mobility switching. It is necessary to fabricate multiphasic polymers to achieve the shape memory effect and control fixing/recovery mechanisms. For example, thermoplastic SMPs such as polymer blends and block copolymers with micro- or nanoscale domain size control require optimum balance between the memory and switching components. Although block copolymers can create well-defined morphologies, polymer synthesis is required in many specialty applications. In contrast, the polymer blends and composites are relatively easy to produce, but achieving uniform domain size dispersion is difficult.

To overcome some of the challenges, multilayer coextrusion processing allowed fabrication of ordered polymeric structures with uniform domain sizes ranging from a few micrometers to a few nanometers [30]. For example, the multilayered structure with alternating layers of PU and polycaprolactone (PCL) were fabricated with varying number of layers and compositions to demonstrate shape memory effect. An example of 50/50 PU/PCL film with a permanent shape, temporary shapes, and a recovered shape is shown in Figure 6.13. A permanent shape – flat film – was fixed into various temporary curved shapes at 70°C (Figure 6.13b)

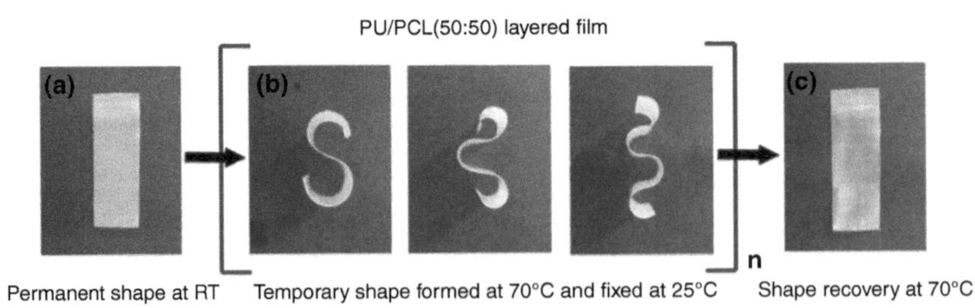

Figure 6.13 Images of PU/PCL (50/50) layered film being shaped multiple times and appearing to recover back to the original permanent shape dimensions.

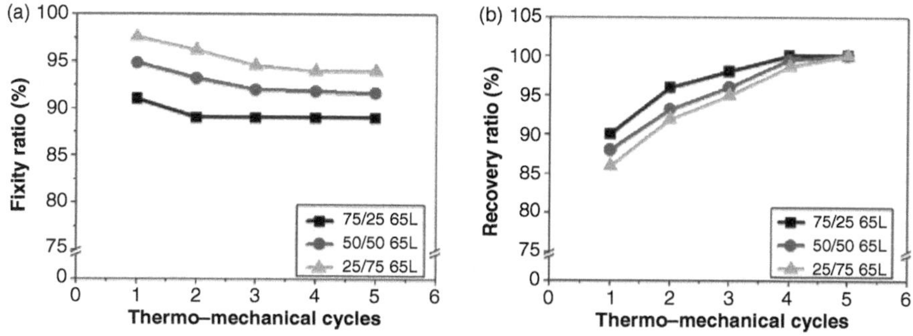

Figure 6.14 (a) Fixity ratio and (b) shape recovery ratios for 65-layered films at different volume compositions.

and maintained at room temperature until further exposure to 70°C causing recovery to its original shape as shown in Figure 6.13c. The shape memory switching temperature for the layered films was between the PCL melting temperature of 60°C and the PU softening temperature of 160°C. Shape fixity and the shape recovery ratios were measured from the stress–strain curves during thermomechanical cycling to quantify the shape memory effect. Shape fixity ratio measures the temporary shape, while shape recovery ratio measures the shape recovery relative to the original permanent shape. In PU/PCL film, higher modulus, and stiffness of PCL as compared to PU at room temperature, controlled the shape fixity, while above PCL melting temperature and below PU softening temperature, PU elasticity contributed to the shape recovery characteristics. The fixity ratio and recovery ratio for the layered films with three different compositions are shown in Figure 6.14. Shape fixity ratio of 91% was obtained in 75/25 PU/PCL film, which increased to 95% in 50/50 and 98% in 25/75 composition films and reduced by 2–3% in subsequent cycles. The shape recovery ratio increased consistently to close to 100% in all layered films. Multiple thermomechanical cycles and conditioning were required to achieve consistent shape memory effect and to reduce PU hysteresis. The morphology analysis of the PU/PCL films showed that the multiple thermomechanical cycles did not affect the layer structure. By optimizing relative compositions of the two materials, it is possible to manipulate the shape memory effect as well as mechanical properties to improve the durability of materials. Furthermore, it is interesting to note that a combination of more than two materials in a multicomponent coextruded system can achieve two temporary shapes within a single composite, also known as triple shape memory effect [30].

6.7 Multilayered Fibers

Polymer fibers have been used commercially for a wide range of applications such as textiles, composite reinforcements, scaffolds, membranes, drug delivery substrates, etc. [31]. Commonly practiced industrial approaches of fabricating polymer fibers can be solvent-based or solvent free. Solvent-based approaches involve solution electrospinning, rotary jet spinning, and solvent casting, while solvent-free techniques involve melt electrospinning and melt blowing.

Electrospinning uses electrical charge to draw submicrometer fibers from polymer solution. In this process, electrical charge is used to fabricate micrometer or submicrometer sized fibers from polymer solution or occasionally from melt [32]. Application of electric potential counteracts the surface tension of the polymer solution thereby facilitating solution droplet stretching to produce a liquid stream. Quick solvent evaporation results in the solidification of polymer fibers. Another solvent-based approach called rotary spinning uses polymer solutions to form fibers, by using centrifugal force [33]. Fiber morphology and structure are controlled in these processes by nozzle geometry, rotation speed, and polymer concentration, among other factors. Fibers in 150 nm to 3 μm diameter have been fabricated using this process. Both these techniques use polymer solutions requiring solvent recovery steps. In addition, low throughput and high-energy consumption are some of the key limitations of these techniques. Recent advancements include gas jet process, which utilizes high velocity expanding gas jets to create polymer solution liquid jets producing polymer nanofibers [34]. With many process modifications to reduce fiber diameter and increase throughput, the solvent handling and recycling remain major disadvantages. Moreover, solvent-based electrospinning cannot be used in most common polyolefins, PP, and PE. In contrast, melt-based polymer fiber fabrication technologies are less commonly used due to significant processing challenges. For example, high viscosity of polymer melts during melt electrospinning technique limits its capabilities to only produce larger fiber diameters. Another technique, melt blowing, is a solvent-free process that can fabricate fibers with diameters < 500 nm [35]. In this technique, polymer is extruded through a die orifice and is blown down to fibers with a hot air jet at the same temperature as the molten polymer. Fundamental understanding of the material selection and process parameters are still in research phase for this technique.

6.7.1 Multilayer Blown Microfibers

Microfibrillation of multilayered films demonstrated the formation of fiber like structures. As discussed in earlier chapters, multilayered films with hundreds of layers consisted of layer thicknesses ranging from a few tens of nanometers to a few micrometers. It is intuitive that the polymer melt consisting of 100s of layers when split into fibers using orifices, it is possible to achieve bi- or multicomponent fibers. An approach to use side-by-side orifices instead of a cast film die produced bicomponent fibers [36]. The filter webs created using the multilayered fibers demonstrated improved filtration efficiency. Various combinations of the two polymers have been demonstrated to create webs with different properties. Such examples include fiber systems of elastomeric and nonelastomeric polymers or heat stable and nonheat stable materials. In these examples, the horizontal layers were kept intact to maintain the layer structure within the fibers. In another approach, the multilayered films were converted into fibrous substrates using fibrillation methods such as hydro-entanglement to separate the adhering horizontal layers [37]. Alternatively, new structures with concentric layers of different polymers were created for optical applications to shift colors [38].

In a recent approach, fiber fabrication using coextrusion process was carried out by using a two dimensional (2D) polymer melt multiplication technique, which created vertical layer structures of two (or more) polymers separated by horizontal layers. The modified process successfully controlled the cross-sectional area of the individual layers. The coextruded structure resembled hundreds of fibers with rectangular cross-section along the extrusion direction. Subsequent separation of all the vertical layers resulted into nonwoven multicomponent or single component fiber webs.

6.7.2 Modified Coextrusion Processing

Multilayer coextrusion process related advances in the processing science have been discussed in detail in Chapter 2. A typical coextrusion process with a two or more polymer processing capabilities is shown in Figure 2.4. A two component system constituted of two extruders connected to two channel feedblocks followed by a series of layer multiplier dies, to increase the number of layers achieving hundreds to a few thousand layers in the final film structure. This process uses a horizontal layer multiplication, where the polymer melt is split vertically at the center and stacked to increase the number of layers parallel to the film surface.

A modified coextrusion process, 2D polymer melt multiplication technique, was designed using conventional extruders and the layer multipliers as schematically shown in Figure 6.15 [39]. Four critical steps in creating a structure relevant for the production of fibers include a two-layer structure as shown in step I. In this step, polymer

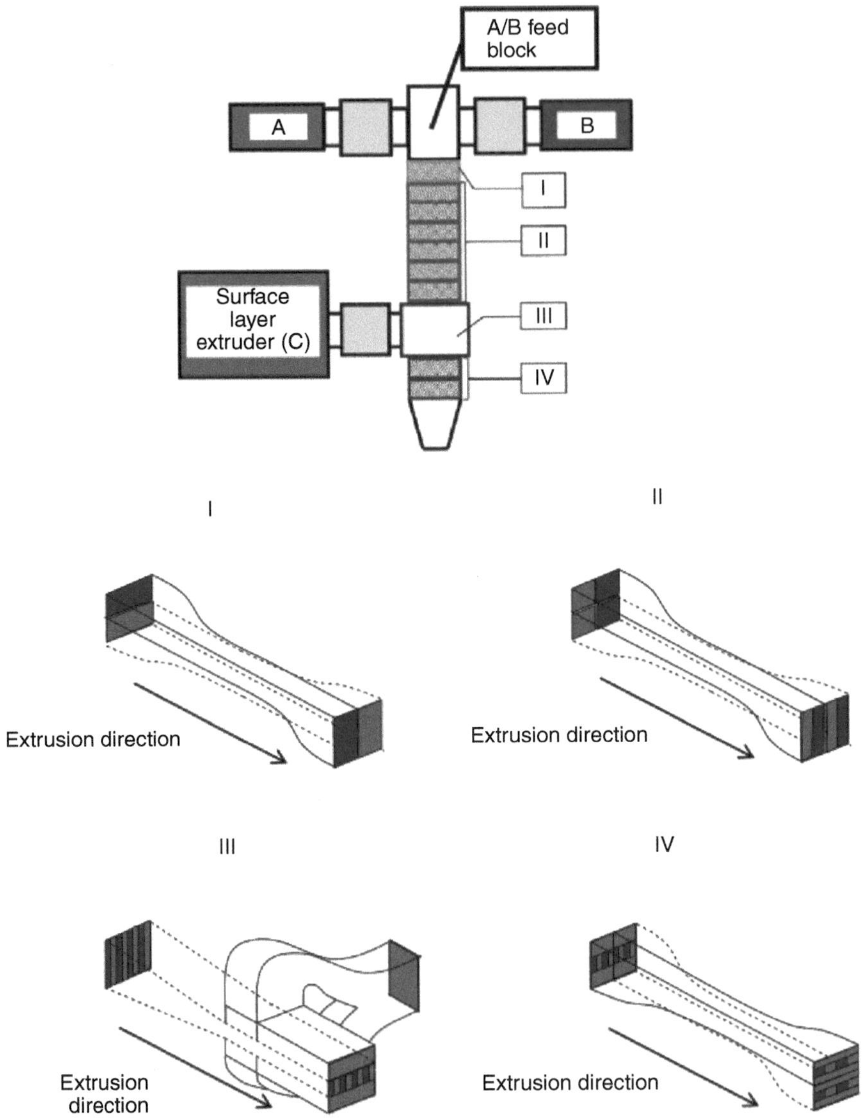

Figure 6.15 Schematics of a coextrusion and 2D multiplication system to produce nanofibers.

melt direction changes from horizontal to vertical in the first multiplier die element. The vertical layer structure was split through a series of multiplying elements (step II) to produce thousands of vertical layers. Each multiplier die cuts the polymer melt in horizontal direction and stacks to align the layers parallel vertically. As the process is repeated multiple times, the number of vertical layers can easily increase to a few thousand. In some examples, few hundred thousand vertical layers were created in this step. The film structure produced during this step can be imagined as thousands of fibers with their thickness in horizontal direction, while the height is in vertical direction. To achieve the true fiber dimension in both dimensions, it is essential to reduce the vertical dimension of the layers and make it comparable with the thickness of the layers. This goal was achieved by adding a third polymer on top and bottom of the vertical layer stack in step III of coextrusion process. In the next, conventional horizontal layer multiplication process cuts the melt flow and stacks the two parts. The extrudate created a 2D structure comprising of vertical layers separated by a sacrificial horizontal layers as shown in step IV. The cross-section images of polymer melt at different stages during the extrusion process, are shown in Figure 6.16. The samples were further processed to separate individual layers to create a polymer fiber matrix. A few examples of fiber matrices created using this technology are discussed later [39].

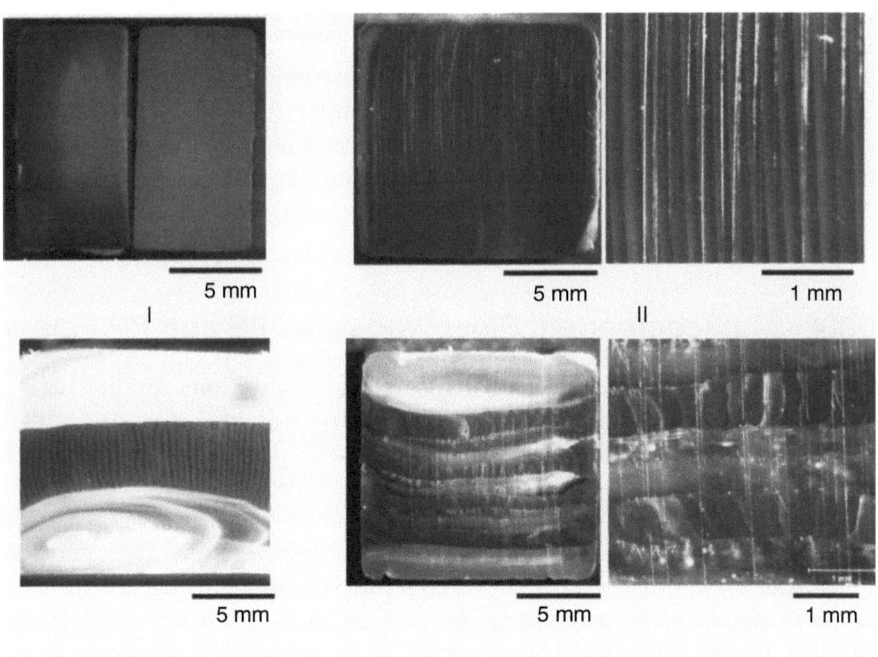

Figure 6.16 Cross-sectional images of samples taken in the extrusion line in each corresponding step described in Figure 6.15.

6.7.3 Multicomponent Fiber Webs – PA6 and PET Fibers

The 2D multilayer coextrusion process was used to produce a layered PA6/PET composite with vertical layer stacks separated by PS layers. PA6 and PET polymers were selected for their weak adhesion with estimated interphase thickness of 3 nm or less. Twelve vertical layer multipliers and four horizontal layer multipliers were used to create a composite structure with 16 stacks of 256 alternating layers of PA6 and PET and each vertical layer stack was separated by PS horizontal layers. The PA6/PET composition of the vertical layers was maintained at 50/50 and the volume ratio of the vertical layers to PS is 2:1. A high-pressure water jet separated the PA6 and PET layers, achieving the composite fiber matrix. The fiber widths ranged from 0.1 μm to 6.6 μm and the thicknesses were in 0.1–0.6 μm. Occasionally, the layers were not delaminated during the separation process resulting into wider fiber size distribution. An orientation process was also incorporated to improve the mechanical properties of the fiber matrix. The fibers from the stretched composites were separated using high pressure water jets to achieve the cross-section dimensions of 0.3–2.4 μm and 0.1–0.4 μm, with more than 80% fibers with cross-sectional dimensions less than 1 × 1 μm. The SEM images of the fibers and fiber size distributions are shown in Figure 6.17. The orientation of fiber composite exhibited almost one order of magnitude improvement in the Young's modulus and tensile strength. The fiber composite created using extruded samples showed Young's modulus and tensile strength of 0.33 ± 0.02 GPa and 54 ± 9 MPa, respectively. In contrast, oriented fibers showed Young's modulus and tensile strength values as high as 6.1 ± 0.2 GPa and 480 ± 21 MPa, respectively [39].

6.7.4 Multicomponent Fiber Webs – LDPE and PP Fibers

Electrospinning techniques have several limitations in producing polyolefin-based nanofibers due to the lack of suitable solvents for dissolving LDPE and PP. However, using these polymers as fibers can be cost-effective due to low resin costs compared to conventional fiber resins. The 2D layer multiplication process demonstrated fabrication of fiber webs from composites with alternating layers of LDPE and PP. Coextruded system of LDPE/PP layers (30/70 composition) comprising of 256 vertical layers separated into 16 stacks by PS horizontal layers. More than 75% of the fibers produced by stretching and water jet separation technique showed width and thicknesses in the range of 0.4–1 μm and 0.2–0.7 μm, respectively. Similar to PA6/PET composite system, the LDPE/PP fiber composites also showed over an order of magnitude

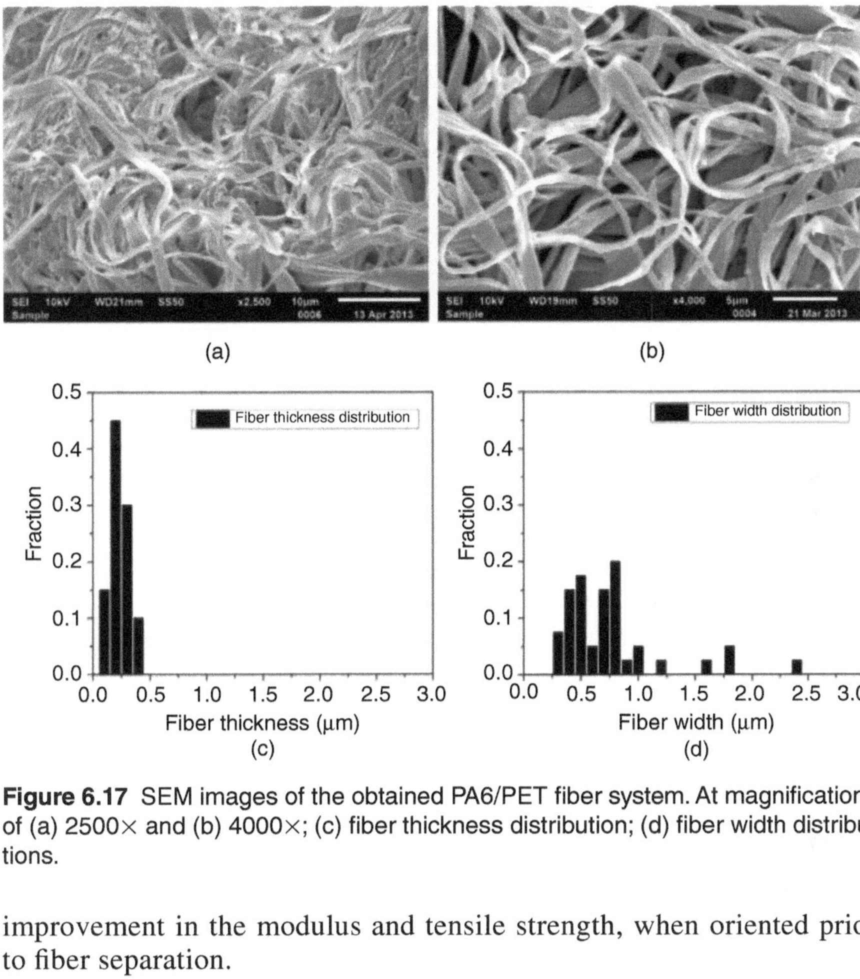

Figure 6.17 SEM images of the obtained PA6/PET fiber system. At magnifications of (a) 2500× and (b) 4000×; (c) fiber thickness distribution; (d) fiber width distributions.

improvement in the modulus and tensile strength, when oriented prior to fiber separation.

6.7.5 Single Component Fiber Webs – PCL Nanofibers

The 2D coextrusion processing approach was further used to coextrude a two-component polymer system with one of the polymers as a water soluble component. In this example, PCL was coextruded with water soluble poly(ethylene oxide) (PEO) to produce four stacks of vertical layered PCL/PEO (256 layers) structure with 10/90 (vol/vol) composition. This structure generated PCL fibers with rectangular cross-section embedded in PEO matrix. After selectively dissolving PEO in water, the composite sample generated a fiber matrix with fiber dimensions in 2–5 μm and 0.4–1.0 μm range. Some thicker fibers were also observed possibly due to incomplete dissolution of PEO. Additionally,

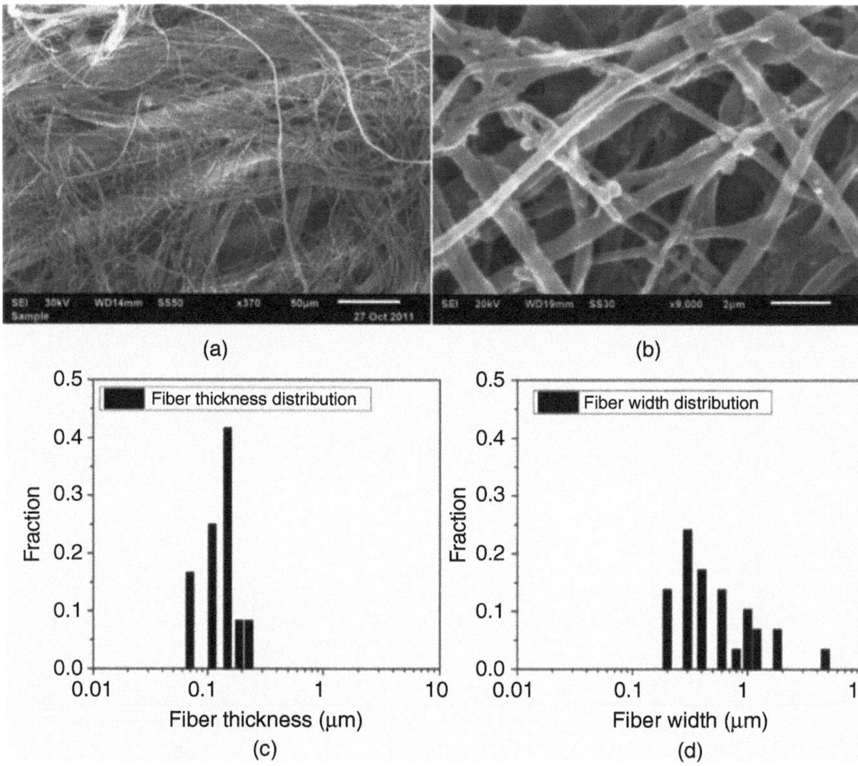

Figure 6.18 SEM images of PCL nanofibers at the magnifications. (a) 370×; (b) 9000×; (c) thickness distribution, and (d) width distribution of the obtained PCL nanofibers.

the composite structure was oriented to reduce the fiber dimension prior to dissolving the PEO. The SEM images revealed decreased fiber dimensions of 0.07–0.3 μm and 0.2–4.7 μm, Figure 6.18. Detailed measurement of the fiber size revealed that more than 90% fibers below 0.2 μm and 1 μm cross-section dimensions, confirming fabrication of PCL nanofibers using coextruded structures. Mechanical properties of the PCL fibers showed Young's modulus increase from 0.2 GPa to 1.6 GPa upon orientation and the tensile strength increase from 16 MPa to 240 MPa.

PCL/PEO system offers unique opportunity to create PCL fiber scaffolds for applications in biomedical field. PCL is commonly used in many applications such as scaffolds for cell and tissue engineering, artificial organs, sutures, and wound dressing due to its biodegradability and biocompatibility [40]. PCL-based scaffolds designed using electrospinning techniques suffer from aforementioned drawbacks as well as weak mechanical properties. A 2D coextrusion processing technique demonstrated the creation of PCL fiber matrix with enhanced mechanical properties. The coextrusion process-based PCL fiber matrix also created surface-immobilized gradients on the fibers by photochemical patterning techniques [41]. Further functionalization of the fiber surfaces was demonstrated as cell-seeding scaffolds [42].

As demonstrated in these examples, the multilayer composite-based fiber fabrication techniques can be used for a wide range of applications. Ability to combine different types of thermoplastic polymers and the process flexibility allow the combination of different types of polymers, which cannot be processed otherwise by conventional fabrication techniques [39].

6.8 Summary

Novel properties of multilayered films and composites have been reviewed in this chapter. The flexibility of the coextrusion technique to process a wide range of polymers into different formulations and layered structure morphology demonstrated a wide spectrum of properties. (a) use of multilayered structures as blends and composites for molding processes produced polymer droplets and the microplatelets. Microplatelets containing composites demonstrated improved gas barrier properties. (b) Different types of fillers in the layers showed ability to create brick-wall type microstructures in the multilayer composites with enhanced mechanical properties. (c) Multilayered foam-film structures revealed improved foaming capabilities and enhanced mechanical properties in polyolefin-based foams. (d) Novel melt die designs allowed the fabrication of gradient structures with a possibility to create custom layer structures for optical applications. This approach was further extended to control the mechanical properties as well as produce gradient foams. (e) The inorganic fillers and nucleating agents in layers allowed formation of porous layers under strain and created gas separation membrane composites for MAP applications when combined with highly selective polymers as layers. (f) Alternatively, layered films with alternating plastic and elastic layers created corrugated structures as breathable films. (g) Multilayer films also produced SMPs by combining two polymers with different stiffness, thermal, and elastomeric properties. (h) Another approach of coextrusion process modification created polymer fiber webs with fiber thicknesses down to a few hundred nanometers.

References

[1] (a) D.R. Paul, C.B. Bucknall, Formulation, Polymer Blends1A Wiley-Interscience Publication, NY, (2000). (b) D.R. Paul, C.B. Bucknall, Performance, Polymer Blends1A Wiley-Interscience Publication, NY, (2000).
[2] T.E. Bernal-Lara, R.Y.F. Liu, A. Hiltner, E. Baer, Polymer 46 (2005) 3043–3055.
[3] Y. Jin, A. Hiltner, E. Baer, J. Polym. Sci. B 45 (2007) 1138–1151.
[4] D.S. Langhe, J.K. Keum, A. Hiltner, E. Baer, J. Polym. Sci. B 49 (2) (2011) 159–171.
[5] (a) D. Jarus, A. Hiltner, E. Baer, Polym. Eng. Sci. 41 (12) (2001) 2162–2172. (b) D. Jarus, A. Hiltner, E. Baer, Polymer 43 (2002) 2401–2408.
[6] (a) Y. Wang, H.W. Milliman, J.R. Johnson III, D.M. Connor, N.A. Mehl, D.A. Schiraldi, Polymer 52 (13) (2011) 2939–2946. (b) Y. Wang, H.W. Milliman, D.M. Connor, N.A. Mehl, D.A. Schiraldi, Polymer 52 (14) (2011) 3226–3233.
[7] (a) M. Gupta, Y. Linn, T. Deans, A. Crosby, E. Baer, A. Hiltner, D.A. Schiraldi, Polymer 50 (2) (2009) 598–604. (b) M. Gupta, Y. Linn, T. Deans, E. Baer, A. Hiltner, D.A. Schiraldi, Macromolecules 43 (9) (2010) 4230–4239.
[8] X. Li, G.B. McKenna, G. Miquelard-Garnier, A. Guinault, C. Sollogoub, G. Regnier, A. Rozanski, Polymer 55 (1) (2013) 248–257.
[9] C.A. Cruz, US Patent No. 8,679,290, 2014.
[10] S. Nazarenko, M. Dennison, T. Schuman, E.V. Stepnov, A. Hiltner, E. Baer, J. Appl. Polym. Sci. 73 (14) (1999) 2877-2885.
[11] (a) N.J. Mills, Polymer Foams Handbook- Engineering and Biomechanics Applications and Design Guide, Butterworth-Heinemann, Burlington, MA, 2007. (b) D. Klempner, V. Sendijarevic, Handbook of Polymeric Foams and Foam Technology, Second ed, Hanser, Munich, (2004).
[12] A.P. Ranade, Structure–Property Relationships in Various Layered Polymeric Systems, Chapter 2, PhD diss., Case Western Reserve University, United States, 2007.
[13] M.A. Barger, D. Bland, M.H. Mazor, E. Baer, J. Dooley, J.A. Garcia, WO 2,008,008,875, A2/US7,993,739, 2011.
[14] A.P. Ranade, A. Hiltner, E. Baer, D.G. Bland, J. Cell. Plast. 40 (2004) 497–507.
[15] H. Sha, I.R. Harrison, J. Poly. Sci. B 30 (1992) 915.
[16] T. Zhao, M. Yang, H. Wu, S. Guo, X. Sun, W. Liang, Mater. Lett. 139 (2015) 275–278.
[17] (a) Rodriguez-Aguilera, J.C. Oliveira, Food Eng. Rev. 1 (2009) 66–83. (b) B.S. Kirland, R. Clarkke, D.R. Paul, J. Membr. Sci. 324 (2008) 119–127. (c) T.C. Mercel, H. Lin, X. Wei, R. Baker, J. Membr. Sci. 359 (2010) 126–139. (d) T. Al-Ati, J.H. Hotchkiss, J. Agr. Food Chem. 51 (2003) 4133–4138.
[18] I.M. Hutten, Handbook of Nonwoven Filter Media, Elsevier, Oxford, United Kingdom, 2007.

[19] S.R. Armstrong, Novel Applications of Co-extruded Multilayer Polymeric Films, Ph.D. Thesis, 2013.
[20] G.T. Offord, S.R. Armstrong, B.D. Freeman, E. Baer, A. Hiltner, D.R. Paul, Polymer 55 (2014) 1259–1266.
[21] A. Exama, J. Arul, R.W. Lencki, L.Z. Lee, C. Toupin, J. Food Sci. 58 (6) (1993) 1365–1370.
[22] V.A. Topolkaraev, US Patent No. 7,303,642, 2007.
[23] A. Yahiaoui, C.F. Thomaschefsky, T. Chiu, US Patent No. 6,417,154, 2002.
[24] (a) A.R. Parker, J. Opt. A 2 (2000) R15–R28. (b) A.R. Parker, D.R. McKenzie, M.C. Large, J. Exp. Bio. 201 (1998) 1206. (c) A.R. Parker, N. Martini, Opt. Laser Tech. 38 (2006) 315.
[25] (a) W.J. Schrenk, US Patent No. 3,773,882, 1973. (b) W.J. Schrenk, US Patent No. 3,884,606, 1975.
[26] M. Ponting, T.M. Burt, L.T.J. Korley, J. Andrews, A. Hiltner, E. Baer, Ind. Eng. Chem. Res. 49 (2010) 12111–12118.
[27] T. Kazmierczak, H. Song, A. Hiltner, E. Baer, Macromol. Rapid Comm. 28 (2007) 2210–2216.
[28] (a) C. Wischke, A.T. Neffe, S. Steur, A. Lendlein, J. Controlled Release 138 (3) (2009) 243–250. (b) M. Behl, M.Y. Razzaq, A. Lendlein, Adv. Mater. 22 (2010) 3388–3410. (c) P.T. Mather, Z. Luo, I.A. Rousseau, Annu. Rev. Mater. Res. 39 (2009) 445–471.
[29] (a) Q. Meng, J. Hu, Composites A 40 (2009) 1661–1672. (b) C. Liu, H. Qin, P.T. Mather, J. Mater. Chem. 17 (2007) 1543–1558. (c) D.C. Lagoudas, Shape Memory Alloys Modeling and Engineering Applications, Springer, US, 2008.
[30] J. Du, S.R. Armstrong, E. Baer, Polymer 54 (2013) 5399–5407.
[31] (a) M.M. Bergshoef, G.J. Vancso, Adv. Mater. 11 (16) (1999) 1362–1365. (b) S.R. Bhattarai, N. Bhattarai, H.K. Yi, P.H. Hwang, D.I. Cha, H.Y. Kim, Biomaterials 25 (13) (2004) 2595–2602. (c) R. Gopal, S. Kaur, C.Y. Feng, C. Chan, S. Rmakrishna, S. Tabe, J. Membr. Sci. 289 (1-2) (2007) 210–219.
[32] J.H. Wendorff, S. Agarwal, A. Greiner, Electrospinning: Materials, Processing and Applications, John Wiley and Sons Ltd, 2012.
[33] M.R. Badrossamay, H.A. McIlwee, J.A. Goss, K.K. Parker, Nano. Lett. 10 (6) (2010) 2257–2261.
[34] R.E. Benavides, S.C. Jana, D.H. Reneker, ACS Macro Lett. 1 (8) (2012) 1032–1036.
[35] C.J. Ellison, A. Phatak, D.W. Giles, C.W. Macosko, F.S. Bates, Polymer 48 (11) (2007) 3306–3316.
[36] (a) D.L. Krueger, J.F. Dyrud, US Patent No. 4,729,371, 1988. (b) E. Joseph, J.A. Rustad, US Patent No. 5,190,812, 1993. (c) E. Joseph, US Patent No. 5,232,770, 1993. (d) E. Joseph, US Patent No. 5,258,220, 1993.
[37] R.S. Kody, M.A. Perez, J.W. Longabach, K.D. Klepzig, J.M. Sebastian, T.R. Hobbs, M.J. Michel, T.J. Lindquist, R.K. Sura, US Patent No. 6,977,113, 2005.

[38] (a) G.L. Bluem, J.M. Frankel, D.C. Kramlich, R.L. Brott, S.D. Hart, L.E Lorimor, P.R. Fleming, W.J. Kopecky, B.B. Wilson, J.M. Jonza, US Patent No. 8,798,421, 2014. (b) S.D. Hart, K.M. Kotchick, W.J. Kopecky, P.R. Fleming, K.L. Thunhorst, US Patent No. 7,327,923, 2008.
[39] J. Wang, D. Langhe, M. Ponting, G. Wnek, L.T.J. Korley, Polymer 55 (2014) 673–685.
[40] (a) H. Sun, L. Lei, C. Song, X. Cui, P. Wang, Biomaterials 27 (9) (2006) 1735–1740. (b) F. Yang, R. Murugan, S. Wang, S. Ramakrishna, Biomaterials 26 (15) (2005) 2603–2610.
[41] S.-E. Kim, E.C. Harker, A.C. De Leon, R.C. Advincula, J.K. Pokorski, Biomacromolecules 16 (2015) 860–867.
[42] S.-E. Kim, J. Wang, A.M. Jordan, L.T.J. Korley, E. Baer, J.K. Pokorski, ACS Macro Lett. 3 (2014) 585–589.

7 Future Trends

7.1 Introduction

Multilayer coextrusion technology is a single step process that combines two or more polymers into films, sheets, or annular structures with tens, hundreds, or thousands of layers using a combination coextrusion processing technique that utilizes layered feedblocks and/or layer multiplier dies. Typically, polymers with dissimilar properties are coextruded to achieve synergistic effect in property improvements. The multilayered films, sheets, or rods have demonstrated a wide range of enhanced properties, which can be used for variety of applications. Most importantly, the layer thicknesses in the multilayered structures can be easily varied between macro-, micro-, and nanoscale to manipulate the structure and properties. The advantages of multilayer coextrusion processes arise from its ability to create a large number of interfaces, which affects the interfacial interactions to impact the overall material properties significantly. The improved properties of coextruded structures have led to many packaging and novel optical applications.

In this book, many novel properties of micro- and nanolayered structures and advantages over the conventional three-, five-, seven-, or nine-layered coextruded articles have been discussed. Multilayered films ranging from two or three layers to a few hundred layers, and in some cases thousands of layers, with individual layer thicknesses ranging from a few micrometers to a few nanometers have been demonstrated. Recent advances in multilayer blown die film technology have demonstrated translation of the cast film techniques in production scale of annular profiles dies and processing lines. In the recent years, many new innovations and detailed understanding of multilayer film structures at different thickness scales have demonstrated additional opportunities to translate the traditionally cast film technology of micro- or nanolayered film processing, to previously untapped conventional as well as emerging markets, as discussed here.

7.1.1 Existing Markets

Packaging and optical applications remain the two primary multilayer film markets since the early inventions of multilayer coextrusion process and products. Some of the early innovations in 1970s demonstrated layered packaging films for high barrier properties with heat sealability and iridescent films for decorative laminations. Since then, processing innovations

Manufacturing and Novel Applications of Multilayer Polymer Films
http://dx.doi.org/10.1016/B978-0-323-37125-4.00007-1

and structural understanding of confined polymers assisted the development of micro- and nanolayer structures for similar applications. In this section, the role and future of multilayer coextrusion that is relevant to packaging and optical applications has been discussed.

7.2 Conventional Packaging Applications

The coextrusion process is accepted as a major fabrication process in the food and consumer goods packaging industry for its economic and technical advantages. Currently, many conventional film-packaging applications use two to five layered structures, while other flexible packaging applications use up to 17-layered structures. Conventional applications such as trash bags, shrink films, frozen food packaging, and meat packaging are typically two to five layer structures containing polyolefins as a major component. In many examples, the polyolefins are coextruded with EVOHs, nylons, PVDCs, and polyesters to achieve improved barrier properties. For example, one of the earliest coextruded materials was a five layer cast film of PE/EVA/Saran/EVA/PE, where Saran acted as a high gas barrier layer, PE layers provided good heat sealability and EVA was used for its good adhesive properties between PE and Saran [1]. In another example, the ABS/HIPS layer structure showed cost reduction to replace all-ABS parts [2]. Over the past few decades, a wide range of polymers have been developed and utilized to improve the mechanical and chemical properties of coextruded structures demonstrating a fundamental approach in coextrusion processing to combine polymers with different properties. In recent years, layered films with improved thermoforming, shrink-wrapping capabilities and oxygen barrier properties were achieved by separating EVOH polymer into microscale layers [3]. Furthermore, many new resins have been developed to use as tie layer materials to improve adhesion between the incompatible component polymers in coextruded structures.

Over last few years, developments in the feedblock designs such as NanoLayer™ feedblock from Cloeren, a Layer Sequencer Technology by BBS Corporation, or commercial cast film modular layer multiplying feedblocks from Nordson/EDI has enable production of layered structures with polymer confinement at nanoscale. As discussed in Chapter 3, nanolayer films with dramatically improved gas barrier properties can be used in packaging film without significant impact on film production cost. For example, using only 5–10% of PCL material as thin nanolayers in the coextruded structure, more than one order of magnitude improvement in barrier properties of the final product can be achieved [4]. It is noteworthy than the barrier improvement is only possible when the PCL layers are present as confined nanolayers with oriented lamellar morphology. In addition, when the nanolayer structures were coextruded with PP skin layers to create BOPP and BOPET films, the nanoconfinement effect improved the barrier properties without sacrificing the optical and the tear properties [5]. The layer thickness, and hence layer morphology, in the film structure can be optimized to achieve combination of desired effects in the final product. For example, the effect of nanolayer coextrusion on mechanical, adhesion, and optical properties can also be extended to create tougher or reflective films in addition to improved barrier.

Many of the micro- and nanolayer coextrusion innovations can serve a \$150 billion (2012) global plastic packaging market, which includes \$40 billion flexible as well as rigid plastics packaging market [6]. Coextruded structures constitute a large fraction of packaging materials for food products such as snacks, cereals, beverages, and it will show continued growth. The multilayered films are expected to serve a growing poultry and meat products market as the sales of these products have grown 15% from 2008 to 2012, and a similar growth is expected over the next few years [7]. The multilayered films will lead market growth in high barrier films, plastic containers, high abuse shrink films, and other packaging applications. In addition to custom designed properties, increasing the number of layers will allow downgauging the film thicknesses while maintaining the desired properties.

7.2.1 Modified Atmosphere Packaging

Modified atmosphere packaging (MAP) materials have been used to increase the shelf life of perishable products such as meats, fruits, and vegetables. According to the United Nations Environment Program, approximately one-third of the food produced worldwide is wasted. Some of the problems can be alleviated by developing inexpensive MAPs to increase the shelf-life of food products. The package design requires creating the equilibrium between CO_2 and O_2 concentrations around the product and increasing the shelf-life. Historically, the conventional packaging films are designed to improve the barrier properties and not for specified CO_2/O_2 selection. Typically MAP materials are designed with a polymer membrane layer with high selectivity values supported on a porous substrate. These MAP technologies use energy intensive fabrication processes to produce ultrathin membranes with desired permeability and selectivity properties. In the examples discussed in Chapter 6, attempts to coextrude layered polymers with high CO_2/O_2 selectivity demonstrated reasonable selectivity values for potential use as MAPs. Typically, the polymers required for high selectivity are expensive than conventional polymers, but it is possible to use these polymers at low concentrations as micro- or nanolayers to achieve desired selectivity properties thereby reducing the final product cost. The layered membranes and films can be researched to develop new packaging and MAPs materials.

7.2.2 Green and Biodegradable Multilayered Films

Over the past few years, the need to produce packaging materials by minimizing environmental impact has been discussed. Among many alternatives, green polymers and biodegradable polymers have been explored as packaging materials. However, as these polymers are more expensive than many petrochemical based polyolefins and resins, their use is still limited in commercial marketplace. In many cases, green polymers or bio-based polymers cannot solely meet material properties requirement of currently produced blend, layered, or composite products. However, as a major component in a multilayered structure, it is possible to produce layered films with green or biodegradable polymers in combination with commercial thermoplastic materials, as well as achieve synergistic property and/or processing improvements. An example of such coextruded products include layered polyester/thermoplastic starch/polyester films [8]. Micro- and nanolayered packaging materials can also be developed to further achieve improved barriers, mechanical, or other properties. A rapid growth in the production of multilayered films with green and biodegradable polymers is envisioned to develop environment friendly applications.

7.2.3 Multilayer Foam/Film Composites

Foamed multilayer films have been developed as rigid as well as flexible packaging materials. Many advantages of layered foam/film structures include reduced weight, improved thermal, and sound attenuation properties. Multilayered foam/film structures with foam density as low as 0.30 g/cm^3 with foaming efficiency of 90% and significantly uniform cell size in PP, EO, and polyolefin blends have been reviewed in Chapter 6. Commercial blown film lines have been used to develop seven-layered foam/film products for food and medical packaging applications by Macro Engineering and Technology. The product examples included biodegradable PLA foams. The stated advantages suggest foam/film use as meat trays eliminating need for water absorbent pads. Continued development work is expected to expand the applications of film/foam systems. Finally, as the coextruded multilayered films, sheets, or foam composites can be thermoformed, it is obvious that the properties of nanolayer structures such as mechanical and barrier improvement can be extended to thermoformed products as high barrier, biodegradable or green materials.

7.3 Layered Optical Films

Decorative iridescent multilayered films have been in use since the invention of layer multiplication technology. Some of the early and current applications of multilayer optical films include decorative films, gift-wraps, ribbons, wall coverings, etc. The advantages of coextrusion processing and nanolayer structure created by combining polymers with different optical properties are discussed in Chapter 4. In recent years, 3M Corporation inventions to control the light reflection by optimizing the multilayer film formulations developed different types of optical films for electronic devices. Some of these new applications of include security window films and novel LED light bulbs. Additional applications of multilayered films include UV reflectors, IR reflectors, reflective polarizers, and optical filters. Layered optical films can be used for passive lighting in buildings as well as reflective films to reduce heat in automobiles or houses. Recently, the uses of layered films in optoelectronic devices have also been demonstrated.

The polymer optical films are very flexible and offer opportunity to develop optical materials for flexible displays, which has gained significant interest in the recent years. One of the key requirements for optical displays is ultra-high barrier films, with reduced edge diffusion. Typical water vapor transmission rates of the order of less than 10^{-3} $cm^3/m^2/day$ are required for enhanced device lifetime, but many existing high barrier polymers showed significantly lower barrier than this target value. Therefore, it is necessary to develop layered films with high barrier properties for this $60 million barrier film market [9]. Recent research of multilayer films in improving the barrier properties may be extended to optical polymers to create micro- or nanolayer ultrahigh barrier optical polymers. It is also possible to incorporate inorganic fillers at nanoscale to improve the barrier properties, while maintaining desired optical properties through enhanced control over the structural dispersion of additive materials. It is envisioned that optical film market will witness a continued growth level in the micro- and nanolayered film systems product arena, as and when recently introduced architectural film market matures and additional markets get identified.

7.4 Molded and Thermoformable Products

Efforts of the film packaging industry for utilizing multilayer processing to enable film down-gauging, while simultaneously maintaining stiffness and barrier, can be extended to other packaging fabrication techniques.

To create layered structure, attempts to simultaneously injection-mold two polymers to create layer structure, have been demonstrated [10]. In contrast, the coextruded structures (films or strands) can be diced or pelletized to use as blends for molding processes, although structural stability in subsequent processing techniques remains a challenge. To expand the advantages of layered geometries beyond films and sheets, the multilayered structures were used as blends to selectively melt one of the component polymers, whereas the second polymer remained dispersed as a microplatelet. Highly dispersed microplatelets exhibited improved barrier properties in injection and compression molded parts as compared to the melt blends. In addition, inorganic fillers are dispersible in multilayered structures to achieve reinforcement. Multilayered blends can be used directly to produce parison in commercial tubular extrusion for the creation of reinforced or high barrier films. It is possible to explore this approach for plastic bottles in liquid packaging applications. These examples demonstrated the application scope of layered structures as innovative blends. As the micro- and nanolayer coextrusion process becomes widely acceptable, a growth in multilayered structures as blends is expected. It is only natural to extend the material optimization techniques implored by film manufacturers to accomplish similar levels of down gauging and property optimization in thicker sheet and injection molded products.

7.5 Annular Coextrusion Technology

Annular dies have been developed and demonstrated on conventional cast and blown films, which constituted up to tens of layers in some cases. However, to create hundreds of layers in tubular dies, complex engineering that ensures concentric layer morphology across the circular manifold is required. In a recent invention, Dooley et al. at Dow Chemical Company, developed multilayer annular profile with a large number of layers using a modified cross-head style blown film die [11]. Thereafter, 114-layered films with 108 core layers and encapsulation layers showed layer integrity at die entry and overlap regions. Other innovations by BBS Corporation developed a layer sequencer technology, which uses modular disk dies stacked together to create nanolayered blown film [12]. As many as 80 metal disks in a die design to create 77-layered structure, have been successfully developed. These dies with layer number flexibility from 25 to 200 are commercially available with high throughput rates. Other annular coextrusion dies, which are available for tubular coextrusion, are developed by Macro Engineering and Technology Inc [13]. These die design advancements of tubular coextrusion with more than 100 layers, can possibly create multilayer parisons and blow-molded parts with micro- and nano-layered structure. Despite complex design requirements, continuous development work in this area will be advantageous in many applications. For examples, using tubular coextrusion process, it is possible to create multilayer parison for blow-molded bottles with improved barrier and mechanical properties. Some early work involved fabrication of blow-molded bottles with few layers [14]. Multilayered pipes for plumbing applications are currently in production [15]. As the processing of annular micro- and nanolayered structures continues to evolve, it is only natural for researchers to extend enhanced toughness findings of cooperative nanolayered material failure mechanisms, that is, cooperative sheer banding and crazing at layer interfaces, to increase the toughness and lifetime of submicron layered piping. Continued investigation in the tubular die and feedblock technology is envisioned.

7.6 Future Markets

Besides packaging and optical film markets, many key advantages and properties of the multilayered films are discussed in this book. The novel properties of layered films allowed exploration of new applications as discussed.

7.6.1 Dielectric Films

Multilayered films with high energy density, high breakdown strength, and relatively low losses were developed by coextruding polymers having different dielectric properties. As discussed in Chapter 5, several publications have demonstrated advantages of layered structure over many current state-of-the-art dielectric polymers such as BOPP and PET. Recently first generation of capacitor prototypes with improved temperature performance have also been developed using multilayer film coextrusion [16]. With more market acceptance and prototype demonstrations, this technology is expected to grow rapidly.

7.6.2 Passive Sensors

As the multilayer film properties can be tuned by optimizing the layer thicknesses and selecting appropriate refractive indices, it is possible to develop multilayer films as passive sensors. For example, narrowband one-dimensional photonic crystals of multilayered films showed changes in transmission spectra under strain [17]. These films can be used as strain sensors or as pressure gauges. Furthermore, the organic dyes can be added to achieve color change in the layered films under external stimuli such as temperature. The layer film sensors can be cost effective products for use as anticounterfeit techniques.

7.6.3 Porous Structures

Multilayered films with various configurations have been used to create microporous structures, as discussed in Chapter 6. Porosity can be created by stretching plastic/elastic-layered films to create corrugated structures or by stretching filled layered structures. The porous materials can be used as absorbent materials in packaging and healthcare applications. Corrugated structures demonstrated advantages as breathable materials, more tactile, and extensible as compared to earlier innovations for personal care products [18]. Layered porous materials can be used as support layers for membrane separation technologies and MAPs.

7.6.4 Shape Memory Composites

Multilayer films have shown promise as shape memory polymers (SMPs), which is a class of materials capable of shape change under external stimuli like temperature, strain, etc. Conventionally, SMPs are custom synthesized to create multiphase polymer with different types of domains. Micro- and nanolayered films fabricated using commercial polymers showed shape memory effect without the need of custom designed materials. The SMPs can be produced cost effectively as passive sensors, actuators, or medical applications. Although multilayered SMP research area is in its infancy, it can be explored to produce commercially scalable shape memory composites for commodity as well as specialty applications.

7.6.5 Fiber Webs and Scaffolds

Different approaches to create fibers and webs have been demonstrated in coextruded structures. For example, the individual layers from layered structures can be separated to create fiber webs or layered melt can be extruded through dies with multiple orifices to create bicomponent fibers, as discussed in Chapter 6. Microlayered structures have been developed to create nonwoven fiber webs as wipe materials, masks, or substrates. Layered systems have been investigated to improve dyeability, electrification, and hydrophilic/hydrophobic properties over single component fibers [19]. Very recently, a "two-directional" layer multiplication was invented at Case Western Reserve University to create fiber webs from unique layer patterns [20]. Examples discussed in this book demonstrate the scalability of coextrusion technique and the ability to compete with conventional fiber production techniques such as electrospinning, melt spinning, etc. As the coextrusion process utilizes wide range of commercial polymers, development of fiber-grade specialty polymers, as required for electrospinning and melt spinning techniques, is not necessary. Coextrusion technique as a route for fiber web production can develop rapidly with the expansion of application oriented material catalog. It is expected that further process optimization will lead to development of fiber webs for applications such as personal care, filtration, and medical products. This approach has also demonstrated its feasibility in cell culture applications using PCL nanofiber web.

7.7 Summary

The coextrusion technology is one of the most important plastic processing techniques, which combines polymers with diverse properties and achieves synergistic effect. Over the last few decades, the coextrusion technique has advanced from two or three layer films to films containing hundreds of layers with vastly improved structure–property relationship knowledge. As the polymer structures approached nanolayer thicknesses, dramatic changes in a wide range of properties were observed, as reviewed in this book. With increased acceptance in the commercial marketplace, many conventional and new applications of coextruded products have been realized. The packaging and optical films will continue to be a driving factor for multilayer film markets. Recent process advancements in creating novel structures such as SMPs and fiber webs will extend the use of coextrusion processes in new markets.

References

[1] J. Eichhorn, Proc. Annu. Nat. Packaging Forum 28 (4) (1966) 242.
[2] W.J. Schrenk, T. ALfrey Jr., in: D.R. Paul, S. Newman (Eds.), Polymer Blends, vol. 2, Academic Press, New York, 1978, pp. 162.
[3] Microlayer films: new uses for hundreds of layers. Plastics Technology Magazine, March Issue, 2006.
[4] M. Ponting, Y. Lin, J.K. Keum, A. Hiltner, E. Baer, Macromolecules 43 (2010) 8619–8627.
[5] Y. Lin, A. Hiltner, E. Baer, Polymer 51 (24) (2010) 5807–5814.
[6] A. Page, Pira Futures Forecast – plastic packaging, Pira Packaging Summit, Nice, France, 2011.

[7] Meat and Poultry Trends in the US Report, Packaged Facts, 2013.
[8] L. Wang, R.L. Shogren, C. Carriere, Polym. Eng. Sci. 40 (2) (2000) 499–506.
[9] H. Zervos, Barrier Layers for Flexible Electronics 2015–2025: Technologies, Markets, Forecasts, IDTechEx, Cambridge, United Kingdom, 2014.
[10] UK Patent No. 1,156,217, 1969.
[11] J. Dooley, J.M. Robacki, M.A. Barger, R.E. Wrisley, S.L. Crabtree, C.L. Pavlicek, US Patent No. 2010/0,215,879, 2010.
[12] (a) H.G. Schirmer, US Patent No. 6,413,595 B1. (b) H.G. Schirmer, US Patent No. 8,870,561.
[13] (a) M. Planeta, S.M. Sagar, US Patent No. 7,097,441. (b) M. Planeta, S.M. Sagar, W. Song, US Patent No. 5,690,972 A.
[14] J. Baird, C.H. Japikse, S.F. Quigley, E.J. Simpson, US Patent No. 4,846,359 A, 1989.
[15] A.M. Borealis and Bourge Pte Ltd, Multilayered Pipes – Solutions for utility piping systems. Available from: http://www.bourge.com/industrysolutions
[16] D. Langhe, M. Ponting, SPE ANTEC, Orlando, 2015.
[17] T. Kazmierczak, H. Song, A. Hiltner, E. Baer, Macromol. Rapid Comm. 28 (2007) 2210–2216.
[18] V.A. Topolkaraev, US Patent No. 7,303,642, 2007.
[19] E. Joseph, US Patent No. 5,232,770, 1993.
[20] J. Wang, D. Langhe, M. Ponting, G. Wnek, L.T.J. Korley, Polymer 55 (2014) 673–685.

Index

FSC
www.fsc.org
MIX
Papier | Fördert gute Waldnutzung
FSC® C083411

Kolibri 360 GmbH
Ferdinand-Jühlke-Straße 7
99095 Erfurt, Deutschland
produktsicherheit@kolibri360.de